Advanced Nanocarbon Materials

Advanced Nanocarbon Materials

Applications for Health Care

Edited by
Sarika Verma, Raju Khan, and Avanish Kumar Srivastava

CRC Press is an imprint of the
Taylor & Francis Group, an **informa** business

First edition published 2022
by CRC Press
6000 Broken Sound Parkway NW, Suite 300, Boca Raton, FL 33487-2742

and by CRC Press
4 Park Square, Milton Park, Abingdon, Oxon, OX14 4RN
CRC Press is an imprint of Taylor & Francis Group, LLC

ISBN: 978-0-367-62054-7 (hbk)
ISBN: 978-0-367-62783-6 (pbk)
ISBN: 978-1-003-11078-1 (ebk)

DOI: 10.1201/9781003110781

Typeset in Times New Roman
by Deanta Global Publishing Services, Chennai, India

Contents

Editor Biographies

Dr Sarika Verma is currently working as a Principal Scientist at the CSIR-Advanced Materials and Processes Research Institute (AMPRI), in Bhopal, India. She has more than 14 years of diverse experience in the area of research and teaching in CSIR and other academic Institutes.

Dr Raju Khan is currently working as a Principal Scientist at the CSIR-Advanced Materials and Processes Research Institute (AMPRI), in Bhopal, India. He has more than 14 years' experience in the field of electrochemical/ fluorescence biosensors and microfluidics.

Dr Avanish Kumar Srivastava is the Director of CSIR-Advanced Materials and Processes Research Institute (AMPRI), in Bhopal, India. He received his master's degree in physics from the Indian Institute of Technology (UOR), Roorkee, his M. Tech. in Materials Science from the Indian Institute of Technology, Kanpur, and his doctorate in metallurgy from the Indian Institute of Science, Bengaluru. He serves as an expert for nanoscale measurements and instrumentation at ISO and BIS. He was instrumental in establishing the high-resolution electron microscopy and spectroscopy facilities at CSIR-NPL, New Delhi, and AMPRI, Bhopal.

Contributors

L. Athira
Minerals Section, Materials Science and Technology Division
CSIR–National Institute for Interdisciplinary Science and Technology (CSIR-NIIST) Council of Scientific & Industrial Research
Kerala, India

Suresh Bandi
Department of Metallurgical & Materials Engineering
Visvesvaraya National Institute of Technology
Nagpur, India

Roberta Bussamara
Institute of Chemistry
Federal University of Rio Grande do Sul
Porto Alegre-RS, Brazil

Rashmi Chowdhary
Department of Biochemistry
All-India Institute of Medical Sciences (AIIMS)
Bhopal, India

Andrea A. H. da Rocha
Institute of Chemistry
Federal University of Rio Grande do Sul
Porto Alegre-RS, Brazil

Manali Datta
Amity Institute of Biotechnology
Amity University Rajasthan
Rajasthan, India

Umesh K. Dwivedi
Amity School of Applied Sciences
Amity University Rajasthan
Jaipur, India

Nathália M. Galdino
Institute of Chemistry
Federal University of Rio Grande do Sul
Porto Alegre-RS, Brazil

H. V. Gangadharappa
JSS College of Pharmacy
JSS Academy of Higher Education and Research
Karnataka, India

Asha P. Johnson
JSS College of Pharmacy
JSS Academy of Higher Education and Research
Karnataka, India

M. A. Mohammed-Aslam
Department of Geology
Central University of Karnataka
Karnataka, India

K. Pramod
College of Pharmaceutical Sciences
Government Medical College
Kerala, India

M. Prasanna
Minerals Section, Materials Science and Technology Division
CSIR-National Institute for Interdisciplinary Science and Technology (CSIR-NIIST)
Council of Scientific & Industrial Research
Kerala, India

S. Ramaswamy
Minerals Section, Materials Science and Technology Division
CSIR-National Institute for Interdisciplinary Science and Technology (CSIR-NIIST) Council of Scientific & Industrial Research
Kerala, India

Deepshikha Rathore
Amity School of Applied Sciences
Amity University Rajasthan
Jaipur, India

R. A. Renjith
Minerals Section, Materials Science and Technology Division
CSIR-National Institute for Interdisciplinary Science and Technology (CSIR-NIIST) Council of Scientific & Industrial Research
Kerala, India

R. G. Rejith
Minerals Section, Materials Science and Technology Division
CSIR-National Institute for Interdisciplinary Science and Technology (CSIR-NIIST) Council of Scientific & Industrial Research
Kerala, India

Chinnu Sabu
College of Pharmaceutical Sciences
Government Medical College
Kerala, India

Renu Sankar
College of Pharmaceutical Sciences
Government Medical College
Kerala, India

Rajesh Saxena
Centre of Excellence in Biotechnology
M.P. Council of Science and Technology
Bhopal, India

Jackson D. Scholten
Institute of Chemistry
Federal University of Rio Grande do Sul
Porto Alegre-RS, Brazil

V. K. Ameena Shirin
College of Pharmaceutical Sciences
Government Medical College
Kerala, India

Ajeet K. Srivastav
Department of Metallurgical & Materials Engineering
Visvesvaraya National Institute of Technology,
Nagpur, India

M. Sundararajan
Minerals Section, Materials Science and Technology Division
CSIR-National Institute for Interdisciplinary Science and Technology (CSIR-NIIST)
Council of Scientific & Industrial Research
Kerala, India

Anita Tilwari
Centre of Excellence in Biotechnology
M.P. Council of Science and Technology
Bhopal, India

M. K. Verma
Basic Medical Research and Premedical Sciences
American University School of Medicine Aruba (AUSOMA)
Aruba, the Netherlands

Sarika Verma
Advanced Materials and Processes Research Institute (AMPRI)
Council of Scientific and Industrial Research
Bhopal, India

1 Carbon Nanotube (CNT)-Mediated Functional Restoration of Human Tissue and Organs

Manali Datta and Umesh K. Dwivedi

CONTENTS

1.1 SYNTHESIS OF CARBON NANOTUBE (CNT) SCAFFOLDS

CNTs are a nano-allotropic form of carbon. The atoms crosslink, forming sheets of six-membered carbon atom rings (graphite) that wrap around themselves to form a cylindrical needle-like structure. They have been broadly classified as single-walled carbon nanotubes (SWCNTs) and multi-walled carbon nanotubes (MWCNTs). Without modifications, CNTs tend to be hydrophobic, so that surface modifications become essential to enhance their functionality in *in-vivo* systems.

Functionalization of the CNTs may be done either covalently or non-covalently (Figure 1.1). Depending on the type of functionalization, modulation of the CNT physicochemical properties occurs, which are furthermore exploited for the regeneration of various tissues and organs.

1.2 SCAFFOLDS AND ORGAN REGENERATION

The use of CNT scaffolds is an application of a phenomenon known as biomimetics. Biomimetics encompasses designs inspired by or derived from living organisms. This up-and-coming specialized area is a direct result and application of converging technologies, a mish-mash of engineering and biological sciences. Nanomaterials

DOI: 10.1201/9781003110781-1

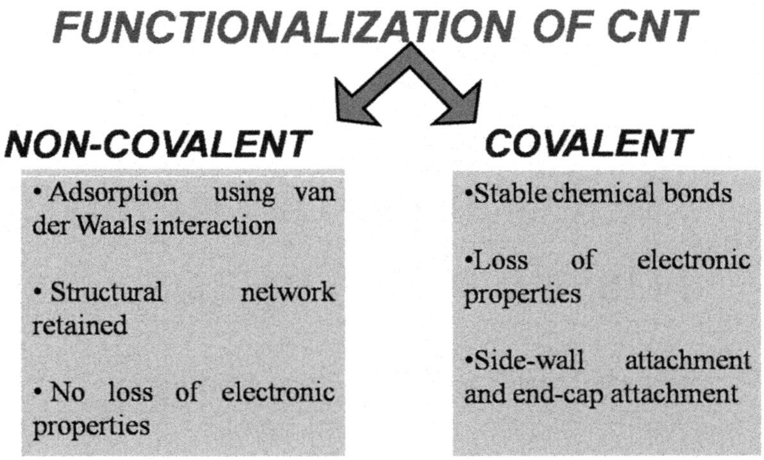

FUNCTIONALIZATION OF CNT

NON-COVALENT
- Adsorption using van der Waals interaction
- Structural network retained
- No loss of electronic properties

COVALENT
- Stable chemical bonds
- Loss of electronic properties
- Side-wall attachment and end-cap attachment

FIGURE 1.1 Different mechanisms of functionalizing CNTs and associated properties.

and nanodevices are thus developed by mimicking natural processes and biological surfaces. Properties of biological materials and surfaces result from a complex interplay between morphology and physicochemical properties. Structural and functional specificity at each hierarchical dimension, from the nano-scale to the macro-scale, determines the efficacy of the biological matrix. The slightest deviation of properties at any level of hierarchy may result in many different functional versions. In this chapter, we will discuss the different types of matrices, the tempering of properties of CNTs, and how modulations enable the CNT scaffolds to act as the perfect template for organ restorations.

1.3 ROLE OF CNTS IN HEART REGENERATION

Cardiac muscles constitute the contractile, functional part of the heart. Highly coordinated contractions enable cardiac muscle, underlined by Purkinje fibres, to act as a conduit for wave-like patterns and thus to act as a pump. Myocardial infarction (MI), an abnormal deformity, results from blood flow restriction to the heart, causing damage to the heart muscle. Subsequent chronic inflammation leads to cardiac remodelling, resulting in loss of elasticity and contractile properties, and leading to tissue damage and heart failure. Heart muscles cannot regenerate themselves and hence represent a perfect target for designing biomimetic scaffolds. Cardiac tissue engineering efforts target the requisite properties: flexibility, electrical conductance, and contractility. Some of the major disadvantages of available scaffold materials are electrical insulation at biocompatible frequencies, the absence of a nanofibrous layout at the sub-micron scale, and lack of mechanical strength. CNT-based biomimetic scaffolds (CNBS) overcome all these disadvantages and show the ability to promote cardiomyocyte proliferation, maturation, and long-term survival, thus upgrading the effectiveness of myocardial engineering.

In the absence of the optimal electro-transducing capability of the biomimetic nanocomposites, loss in signal transduction results in heightened arrhythmias. Thus, one of the first requisites for recruiting CNBS involved standardizing excitation conduction velocities to approximately 22 ± 9 cm/s. Single-walled nanotubes at minimal concentrations (<100 ppm) were dispersed in a gelatin/chitosan solution to form hydrogels (Pok 2014). Using Di-8-ANEPPS, a voltage-sensitive fluorescent dye to gauge the efficacy of the conduction velocity of the CNTs, rat myocardiocytes were grown on SWCNT-impregnated hydrogel as a substratum and were found to achieve a velocity equivalent to 23 cm/s. On the other hand, bioprinted CNT-based hybrid implants, containing alginate and methylcobalamin, provided the hybrid implants with improved stiffness, conductivity, and cellular response.

Another CNBS was designed by introducing CNTs in poly(octamethylene maleate (anhydride) 1,2,4-butanetricarboxylate), henceforth known as 124 polymers. The scaffold formed had optimal electrical conductivity and structural integrity. Variation in the CNT concentration in the polymer resulted in an increase in surface moduli with a simultaneous decrease in bulk moduli. Placed in an aqueous environment, 124 polymers exhibited increased swelling, indicating an improved structural efficacy in a physiological environment. With 0.5% CNT content, the excitation threshold of the CNBS was found to be 3.6 ± 0.8 V/cm, confirming its potential to act as a versatile CNBS for cardiac tissue engineering (Pok et al. 2013, 2014).

A cardiopatch combining bacterial cellulose and SWCNTs was synthesized and its efficacy as an adduct was checked by attaching it to the epicardium before and after surgical intervention. Electro-mapping was used to assess the functioning of normal and disrupted epicardium, both pre- and post-application of patch. Three-dimensional printable carbon nanotube ink, complexed on bacterial nanocellulose, was (1) expressible through 3-dimensional printer nozzles, (2) electrically conductive, (3) flexible, and (4) stretchable (Pedrotty et al. 2019).

Flexible scaffolds, composed of CNTs embedded in poly (glycerol sebacate): gelatin (PG), resulted in a nanofibre with enhanced electrical properties. Furthermore, increases in the concentration of CNTs (0–1.5%) within the CNT–PG scaffolds tended to improve fibre alignment, toughness, and electrical conductivity, while maintaining the essential contractile property characteristics of cardiomyocytes. Another such conjugation was attempted by electro-spraying multi-walled carbon nanotubes (MWCNTs) on polyurethane nanofibres. In addition to improving the electrical conductivity and tensile strength, the hydrophilicity of the CNT/polyurethane (PU) nanocomposites was enhanced. Cell cytotoxicity studies confirmed the cytocompatibility of CNT/PU nanocomposites in rat and human cells. Several more 2D- and 3D-printed cardiac patches have been synthesized using gelatin methacrylate (CNT/GelMA) or collagen (CNT/Col) (Ahadian et al. 2017, Pedrotty et al. 2019, Shokraei et al. 2019) (Figure 1.2).

Chemically synthesized 3D graphene foam and 2D graphene etched with nickel were other templates used to regenerate cardiomyocytes. Toxicity testing revealed no toxic effects on the human umbilical vein endothelial cells for up to three days. Enhanced adhesion, as well as increased expression of genes involved in muscle contraction and relaxation (troponin-T) and gap junctions (Connexin 43), was observed

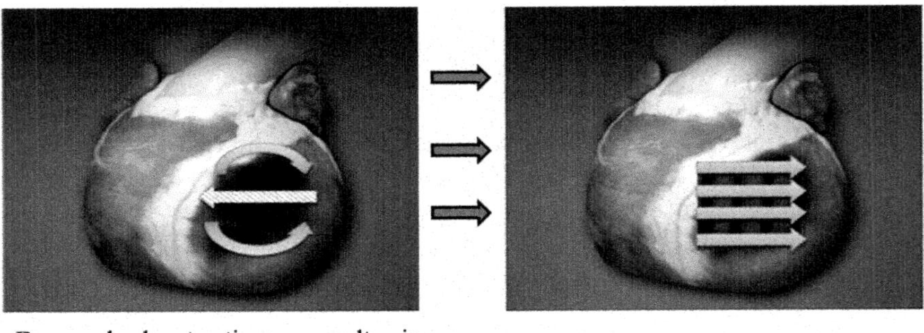

FIGURE 1.2 Schematics of the use of CNTs in overcoming cardiac arrthymia. A CNT-based cardiac patch is being used to allow normal conduction of the cardiac cycle in scarred tissue of the heart (Pedrotty et al. 2019).

from the cardiomyocytes. Thus, this conductive 3D graphene foam, with its large surface area, was established as a promising platform for cardiac tissue engineering (Martinelli et al. 2018).

1.4 ROLE OF CNTS IN BONE REGENERATION

Bone malformations have traditionally been alleviated by medical procedures like autografting, allografting, and tissue transfer techniques. As grafting procedures have risks associated with them, like disease transmission, tissue rejection, and failure to induce bone cell or structure formation, alternatives were investigated to mimic properties characteristic of bone. Carbon nanostructures are promising materials for bone tissue engineering on the grounds of their low weight, and high strength, conductivity, and stability. Synthetic bone substitutes used as alternatives were osteoconductive with minimal risk of disease transmission. Some of the properties which highlighted the potential of CNTs as materials for bone tissue engineering were the possibility of achieving cylindrical shapes of nanoscale diameters, increased aspect ratios, tensile strength $\geq$50 GPa, Young's modulus $\geq$1 TPa, conductivity $\sigma in \geq 107$ S/m, maximum current transmittance $Jin \geq 100$ MA/cm^2, and density $\rho \leq 1600$ kg/m^3 (Keihan et al. 2019).

By themselves, CNTs do not have the physicochemical requirements to support the differentiation of osteocytes. Hence, CNTs are generally coated with polymers to enhance their ability to act as a regenerating bone platform. Initially, SWCNTs were used in combination with poly(lactic-co-glycolic acid), or PLGA, a US Food and Drug Administration-approved biocompatible polymer, over 12 weeks *in vivo* in rats, thereby confirming acceptable biocompatibility to allow long-term *in-vivo* usage. In conjunction with SWCNTs, polycaprolactone (PCL), a biocompatible polymer, could also enhance their structural and functional properties, mediating the attachment, proliferation, and differentiation of rat bone marrow-derived mesenchymal stem cells (rMSCs) on the composite. The composite had thinner fibres with improved tensile strength, and greater electrical conductivity with decreased bioactivity and degradation. Enzymatic monitoring of the growth of stem cells indicated that the highest viability was achieved on a PCL/aSWCNT-0.5 scaffold. These results confirmed that SWCNT/PLGA composites, containing a low concentration of SWCNTs, were not generating localized or generalized toxicity. Scientists across the world have used different polymers, such as bacterial cellulose, poly(D,L-lactic acid) (PDLLA), poly(L-lactide-co-glycolide) (PLGA), poly(lactic acid) (PLA), to name but a few, to generate sturdy scaffolds with desired features necessary for bone regeneration (Gupta et al. 2015, Perkins and Naderi 2016).

Hydrogels have been designed to mimic the biological extracellular matrix (ECM), but they lack the mechanical strength required to support tissue constructs. Gelatin methacrylate-based gels impregnated with CNTs form a GelMA hybrid. GelMA provides a suitable platform for spiking the growth of tissues, subsequently leading to three-dimensional (3D) constructs. Balancing the ratio of CNTs in GelMA could alter the hybrid material's mechanical properties, making it suitable for various biomimetic applications (Gorain et al. 2018).

A biodegradable gel consisting of oligo(poly(ethylene glycol) fumarate) (OPF) infused with 2D black phosphorus (BP) nanosheets and embedded in cross-linkable CNT-poly(ethylene glycol)-acrylate (CNTpega) (known as BP-CNTpega-gel) has also been used as an alternative for a bone regeneration platform. BP-CNTpega-gel could induce osteogenic differentiation of osteoblast cell lines. In the rabbit, *in-vivo* studies indicated that, with electric stimulation (100 mV/mm at a frequency of 20 Hz), the expression of essential osteogenic pathway genes was successfully induced, thus enabling *in-situ* rectification of femur defects, vertebral cavities, and spinal fusion (Liu et al. 2020).

1.5 ROLE OF CNTS IN NERVE REGENERATION

Neural regeneration depends on either repair or replacement of nerve cells. In this respect, CNTs have emerged as a promising biomaterial for neural regeneration and retention of neural biological functionality. CNTs have morphological similarities, with the small CNT bundles having dimensions similar to those of dendrites, thus eliciting neural probing, repair, or reconfiguring of neural networks. CNTs can sustain and promote neuronal electrical activity at the cellular level. Various biophysical methods to monitor single-cell electrophysiology, such as electron microscopy and theoretical modelling, indicated that nanotubes could elicit electrical impulses by forming tight contacts with the cell membranes. This enabled the formation of electrical shortcuts between the proximal and distal compartments of the neuron, a phenomenon known as the electrotonic hypothesis. Chemical modification ("biofunctionalization") of CNTs resulted in improved biocompatibility and selectivity for neural regeneration. CNT-patterned scaffolds can be synthesized using techniques like photolithography and chemical vapour deposition (CVD). However, since some scaffolds are based on rigid substrates like glass, aluminium foils, or silicon, they occasionally induce nerve compression, imposing weakening of the nerve conduit.

Literature has reported that amines can promote neuron growth. An amine-functionalized MWCNT–PEGDA polymer was synthesized using stereo-lithography, conferring on it a tunable porous structure. An additional feature of this nano-scaffold was its enhanced electrical conductivity. The scaffold promoted superior neural proliferation and differentiation compared with other templates in use, with the induction of cellular maturity. A simple biphasic pulse stimulation with a 500 μA current induced neuronal maturity. This platform is a classic example of synergy between constituents providing the appropriate niche for neural regeneration (Lee et al. 2018).

Silk has also been used as a template to generate a CNT-based scaffold, by electrospinning. This template could exhibit the combination of mechanical properties, hydrophilicity, biodegradability, and biocompatibility with nerve tissue. Polyurethane (PU)-based PU/silk fibroin–fMWCNT biomaterial could also exhibit neuronal growth and differentiation. PU/silk–fMWCNTs promoted the proliferation of Schwann cells, in addition to differentiation in rat pheochromocytoma (PC12) cells. The fMWCNTs contributed to the electrical conductivity, as well as absorption of sufficient extracellular matrix (ECM) to open up a new perspective on peripheral nerve restoration, using silk as a feasible material for generating tissue engineering platforms (Figure 1.3) (Ahn et al. 2015).

Chitin has also been used extensively for the design of templates for tissue engineering. The chitin composite with MWCNTs ensured that CNTs aligned with the matrix, with an almost negligible risk of detachment and leaching from the surface. A robust synthesis method was formulated, whereby an AC electric field was used to align the MWCNTs within the chitosan matrix. Native chitin–MWCNTs allowed the unidirectional growth of neurons on its surface, thus ensuring the appropriate

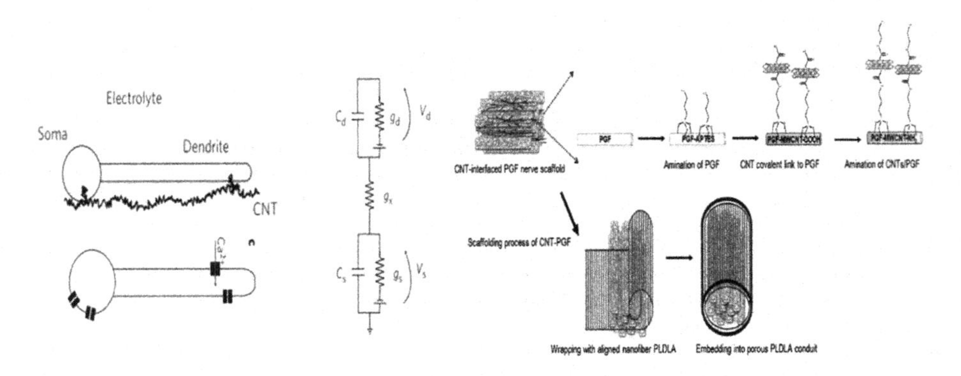

FIGURE 1.3 Intracellular compartments are electrically layered on nanotubes, making contacts and resulting in an electrical shortcut. A two-compartment spiking neuron model could mimic CNT-induced ADP, thus generating a neuronal network. PGFs interfaced with CNTs may be rolled upon a PLDLA electrospun nanofiber mat to resemble a nerve bundle (Ahn et al. 2015).

conductivity of the composite. One of the combinations developed for neural applications involved chitin–MWCNTs embedded in a hydrogel bulk which is stabilized through intermolecular forces like hydrogen bonding, electrostatic, and hydrophobic interactions. The hydrogels had enhanced thermostability, haemocompatibility, and mechanical strength, with increased sustainability *in vivo*. Another functional composite, CNT- embedded hyaluronic acid (HA), was generated by electrospinning. Various biophysical characterizations of this CNT–HA nanofibre confirmed the functionality of the artificial tissue. The electrochemical testing used for functional efficacy was electrical impedance spectroscopy (EIS) and cyclic voltammetry (CV), which confirmed that conductivity and charge storage capacity were significantly enhanced (Cellot et al. 2009, Gupta and Lahiri 2016, Wu et al. 2017).

1.6 ROLE OF CNT IN SKIN REGENERATION AND WOUND HEALING

Wound healing is another medical category which requires tissue regeneration. Although a natural process, health anomalies like diabetes, bedsores, and extensive burns sometimes require external intervention to achieve full recovery. Extracellular matrix (ECM), the natural template which induces the morphogenesis leading to tissue development, contains various biological molecules, including noncollagenous glycoproteins and growth factors. ECM facilitates various activities, such as cell polarization, migration, stabilization, cell contraction, and detachment, to allow formation of a skin. The artificial template is highly hydrated with varying stiffness levels, and needs to mimic ECM to form an ideal biomimetic platform. Hence, most of the platforms generated for skin regeneration consist of composites embedded in hydrogels.

A simple biological dressing consists of a composite of poly(vinyl alcohol) (PVA), CNTs, and a natural skin repair inducer, epidermal growth factor (EGF), achieved by electrospinning. The *in-vitro* safety of the composite was tested using a methyl thiazolyl tetrazolium (MTT) assay on L929 fibroblasts. The results indicated mild toxicity, due to a sustained release of EGF from the patch, thus enabling slow yet steady repair of the injured skin. An *in-vivo* experiment was conducted to observe whether this dressing could improve healing in the model of wounded skin on Sprague–Dawley rats. Gross and histological testing indicated rapid healing, thus confirming that the composite may be used for accelerated wound healing (Liao et al. 2017).

Nano-embedded hydrogels, especially CNT-loaded hydrogels, have been shown to promote cell concentration at and cell migration to the focal point of injury. Hydrogels with chitosan and either SWCNTs or MWCNTs enabled the formation of a matrix with a competent hydrophilic nature capable of effective contact with skin and the formation of consistent tissue. Chitosan incorporated with SWCNTs or MWCNTs has been successfully used to achieve regeneration therapeutics in wound injuries. Microscopic evaluation and semi-quantitative assays confirmed that fibroblasts could contract and assemble the collagen, thus forming a viable tissue without any toxic aftereffects. Studies indicated that such a composite could successfully heal a wound within nine days. Composites containing 1% C-SWCNTs or 5% C-MWCNTs resulted in wounds being drier and minutely sutured. There was an almost 40% increase in collagen deposition in 0.5% C-SWCNTs and a 120% increase upon using 5% C-MWCNTs as compared to the CNT free composites. Another composite consisted of carbon nanotubes, silver nanoparticles (AgNPs), and polyvinyl alcohol, electrospun to generate nanofibres. In addition to its wound healing capability, this composite also demonstrated bactericidal and bacteriostatic activity, further justifying its role as a suitable template for designing sustainable wound dressings. CNTs, in combination with glycidyl methacrylate and chitosan, have also been used as a wound dressing for haemorrhages, haemostasis, and wound healing. The composite provided an ideal template for regeneration of fibroblasts, in addition to skin shape recovery and high blood-absorption capacity (Kittana et al. 2018, Jatoi et al. 2019).

1.7 CONCLUSION

Nanostructured materials based on CNTs have demonstrated unmatched efficacy in terms of tissue regeneration. Over time, the properties of CNT composites have been manipulated to be simultaneously biodegradable and biocompatible for a specific application. Depending on their scaffold constituents, these composites can elicit many responses, depending on cell type, feature size and geometry, and substrate stiffness. Although the advantages of CNT-based scaffolds have been highlighted, these nanostructures have exhibited cytotoxicity in immune cells. CNT based scaffolds attract and activate the alveolar macrophages and human T-lymphocytes, thus aggravating the host immune response. Even though CNT-based composites have shown tremendous potential and even success as templates for regenerative scaffolds, further assessment in clinical trials is essential before they can be certified for human applications.

1.8 REFERENCES

Ahadian, Samad, Locke Davenport Huyer, Mehdi Estili, et al. "Moldable elastomeric polyester-carbon nanotube scaffolds for cardiac tissue engineering." *Acta Biomaterialia* 52 (2017): 81–91.

Ahn, Hong-Sun, Ji-Young Hwang, Min Soo Kim, et al. "Carbon-nanotube-interfaced glass fiber scaffold for regeneration of transected sciatic nerve." *Acta Biomaterialia* 13 (2015): 324–334.

Cellot, Giada, Emanuele Cilia, Sara Cipollone, et al. "Carbon nanotubes might improve neuronal performance by favouring electrical shortcuts." *Nature Nanotechnology* 4, no. 2 (2009): 126–133.

Eivazzadeh-Keihan, Reza, Ali Maleki, Miguel De La Guardia, et al. "Carbon based nanomaterials for tissue engineering of bone: Building new bone on small black scaffolds: A review." *Journal of Advanced Research* 18 (2019): 185–201.

Gorain, Bapi, Hira Choudhury, Manisha Pandey, et al. "Carbon nanotube scaffolds as emerging nanoplatform for myocardial tissue regeneration: A review of recent developments and therapeutic implications." *Biomedicine & Pharmacotherapy* 104 (2018): 496–508.

Gupta, A., T. A. Liberati, S. J. Verhulst, B. J. Main, et al. "Biocompatibility of single-walled carbon nanotube composites for bone regeneration." *Bone & Joint Research* 4, no. 5 (2015): 70–77.

Gupta, Pallavi, and Debrupa Lahiri. "Aligned carbon nanotube containing scaffolds for neural tissue regeneration." *Neural Regeneration Research* 11, no. 7 (2016): 1062.

Jatoi, Abdul Wahab, Hiroshi Ogasawara, Ick Soo Kim, and Qing-Qing Ni. "Polyvinyl alcohol nanofiber based three phase wound dressings for sustained wound healing applications." *Materials Letters* 241 (2019): 168–171.0960374000

Kittana, Naim, Mohyeddin Assali, Hanood Abu-Rass, et al. "Enhancement of wound healing by single-wall/multi-wall carbon nanotubes complexed with chitosan." *International Journal of Nanomedicine* 13 (2018): 7195.

Lee, Se-Jun, Wei Zhu, Margaret Nowicki, et al. "3D printing nano conductive multi-walled carbon nanotube scaffolds for nerve regeneration." *Journal of Neural Engineering* 15, no. 1 (2018): 016018.

Liao, JunLin, Shi Zhong, ShaoHua Wang, et al. "Preparation and properties of a novel carbon nanotubes/poly (vinyl alcohol)/epidermal growth factor composite biological dressing." *Experimental and Therapeutic Medicine* 14, no. 3 (2017): 2341–2348.

Liu, Xifeng, Matthew N. George, Linli Li, et al. "Injectable electrical conductive and phosphate releasing gel with two-dimensional black phosphorus and carbon nanotubes for bone tissue engineering." *ACS Biomaterials Science & Engineering* 6, no. 8 (2020): 4653–4665.

Martinelli, Valentina, Susanna Bosi, Brisa Peña, et al. "3D carbon-nanotube-based composites for cardiac tissue engineering." *ACS Applied Bio Materials* 1, no. 5 (2018): 1530–1537.

Pedrotty, Dawn M., Volodymyr Kuzmenko, Erdem Karabulut, et al. "Three-dimensional printed biopatches with conductive ink facilitate cardiac conduction when applied to disrupted myocardium." *Circulation: Arrhythmia and Electrophysiology* 12, no. 3 (2019): e006920

Perkins, Brian Lee, and Naghmeh Naderi. "Suppl-3, M7: carbon nanostructures in bone tissue engineering." *The Open Orthopaedics Journal* 10 (2016): 877.

Pok, Seokwon, Jackson D. Myers, J. G. Jacot, et al. "A multilayered scaffold of a chitosan and gelatin hydrogel supported by a PCL core for cardiac tissue engineering." *Acta Biomaterialia* 9, no. 3 (2013): 5630–5642.

Pok, Seokwon, Flavia Vitale, Shannon L., et al. "Biocompatible carbon nanotube–chitosan scaffold matching the electrical conductivity of the heart." *ACS Nano* 8, no. 10 (2014): 9822–9832.

Shokraei, Nasim, Shiva Asadpour, Shabnam Shokraei, et al. "Development of electrically conductive hybrid nanofibers based on CNT-polyurethane nanocomposite for cardiac tissue engineering." *Microscopy Research and Technique* 82, no. 8 (2019): 1316–1325.

Wu, Shuangquan, Bo Duan, Ang Lu, et al. "Biocompatible chitin/carbon nanotubes composite hydrogels as neuronal growth substrates." *Carbohydrate Polymers* 174 (2017): 830–840.

2 Advanced Nanocarbon Materials

An Introduction to Their History and Background

Deepshikha Rathore and Umesh K. Dwivedi

CONTENTS

2.1 INTRODUCTION

Carbon is a significant element of this universe for all non-living and living things, such as plants and animals, including humans, because all organic compounds are composed of a carbon matrix. Charcoal is an example of the purest form of carbon material with low oxygen content, which was discovered as far back as 3,750 BC and used in the formation of bronze. Activated charcoal was also used as medicine to cure epilepsy, vertigo, and chlorosis (hypochromic anemia) in ancient times. In 748 CE, almost 800 tons of charcoal were used to cast statues of Buddha in Nara City, Japan. At the beginning of the 16th century, graphite was discovered as a soft, lubricating carbon material and made into pencil lead and carbon black for ink. Later on, graphite electrodes were produced in huge capacity and used in refining metals. In the middle of the 18th century, diamonds came into the picture as the hardest

DOI: 10.1201/9781003110781-2

transparent rock. It was later used to design jewellery after cutting and polishing due to its fascinating optical properties. In contrast to new carbon materials, "charcoal," "graphite," and "diamond" are classic carbon materials (Inagaki 2000).

In addition to these classic materials, some other carbon allotropes have been discovered with various properties and applications, known as first-generation carbon materials. Initially, in the 1970s, Morinobu Endo, a renowned Japanese researcher and the director of the Institute of Carbon Science and Technology at Shinshu University, discovered carbon nanofibers, including hollow tubes. These nanofibers were used extensively in drug delivery, cancer diagnosis, and tissue engineering. In 1985, fullerene, which contains 60 carbon atoms (C_{60}), was discovered by Harold W. Kroto of the United Kingdom and Richard E. Smalley and Robert F. Curl, Jr., of the United States as an allotrope of carbon. Fullerene has been used widely as lubricants, catalysts, and drug delivery agents in the body.

Subsequently, Sumio Iijima discovered new carbon nanotube materials in 1991, which exhibited highly efficient electrical, mechanical, and thermal properties. As a consequence, carbon nanotubes have been extensively used in electronic components, energy storage, electrical, chemical, mechanical, and optical applications. More recently, graphene was discovered by Andre Geim and Kostya Novoselov of the University of Manchester in 2004, exhibiting a beautiful single layer of carbon atoms in a lattice structure. The potential applications of graphene, the thinnest and strongest material in the world, were flexible, thin, and lightweight photonic and electric circuits, solar cells, and various chemical, medical, and industrial developments.

After the first-generation carbon allotropes, many research groups in the past two decades investigated the second-generation carbon materials in the form of functionalized and doped nanocarbons. Moreover, advanced hybridized carbon nanomaterials with enhanced and efficient properties and potential applications were also explored in the past decade. All these nanocarbon materials will be discussed in the next section, based step-by-step on material generation (Figure 2.1).

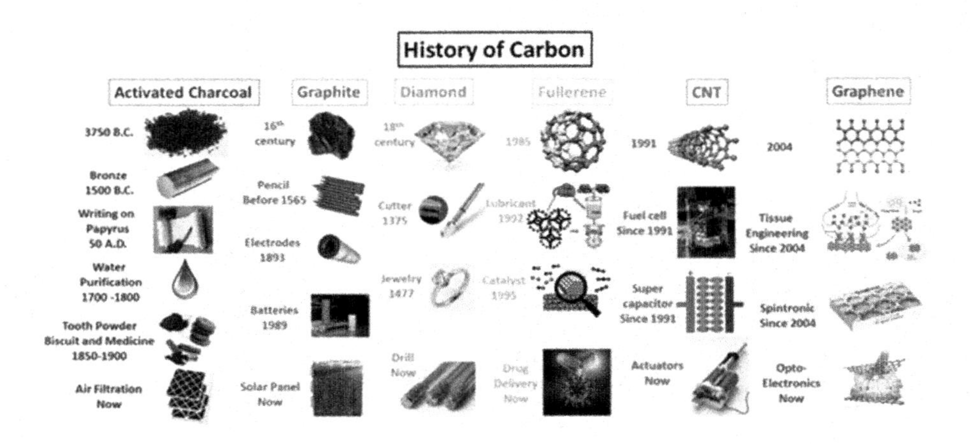

FIGURE 2.1 History of carbon.

2.2 FIRST-GENERATION CARBON NANOMATERIALS

Fullerene (C_{60}), graphene, and carbon nanotubes (CNTs) are attractive nano forms of carbon. The basic features, with various properties and different applications, are discussed in this Section 2.2.

2.2.1 Fullerene (C_{60}) and Its By-products

Several physicists and chemists had envisioned the arrangement of 60 carbon atoms like a soccer ball for many decades. The C_{60} had been named fullerene to commemorate the great architect R. Buckminister Fuller, who designed C_{60} like a geodesic dome. The structure of the C_{60} fullerene molecule is shown in Figure 2.2 (a). It has 20 hexagonal and 12 pentagonal faces organized symmetrically into a ball molecule. These carbon atoms bind to form crystal lattices containing face-centred cubic structures, as illustrated in Figure 2.2 (b). In the C_{60} fullerene molecular lattice, each carbon atom is separated by 1 nm from its nearest neighbouring atoms with weak van der Waals forces. However, a C_{60} fullerene molecule can dissolve in benzene, and on evaporation of the solution, a single crystal of C_{60} can be grown. A total of 26% of the C_{60} cubic lattice is unoccupied. Hence, alkali atoms can be doped very quickly.

Although the C_{60} crystal shows insulating properties, it becomes conducting when potassium is doped in the K_3C_{60} compound. In this way, K atoms become ionized, and each C_{60} molecule gains three electrons. These three electrons can move inside the C_{60} lattice and make it electrically conducting. The alkali atoms occupy

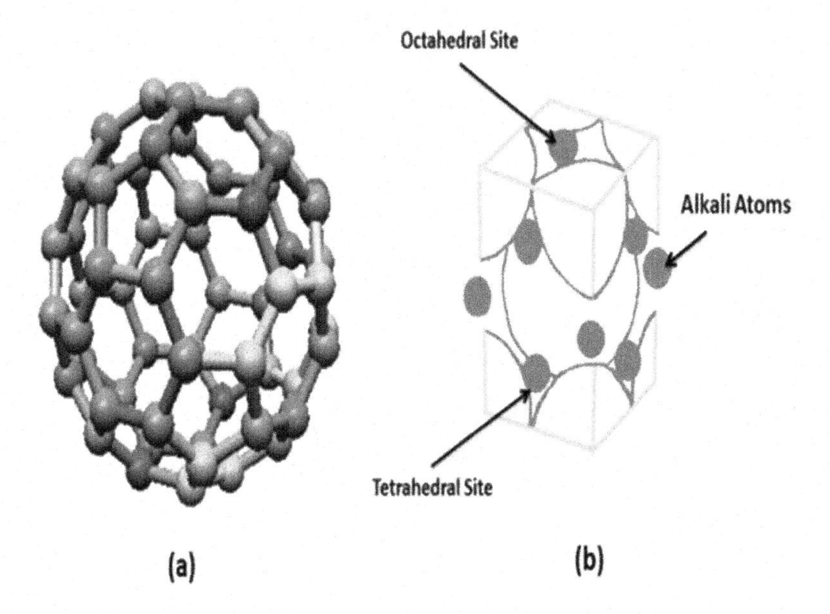

FIGURE 2.2 (a) Structure of C_{60} fullerene molecule (by Nanotube Modular Software) and (b) crystal lattice unit cell of C_{60} fullerene molecules doped with alkali atoms.

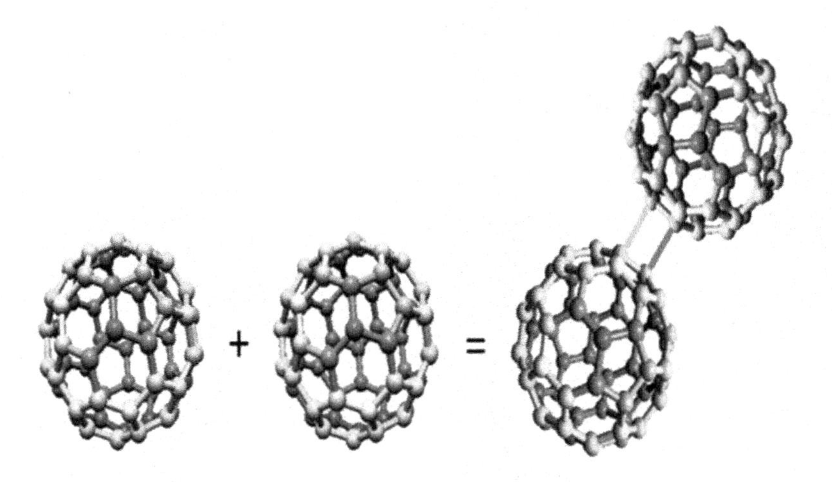

FIGURE 2.3 Structure of C_{60} dimer (by Nanotube Modular Software).

one vacant octahedral and two vacant tetrahedral sites per C_{60} fullerene molecule. Thus, these are surrounded by four and six C_{60} atoms in the tetrahedral and octahedral sites, respectively, as illustrated in Figure 2.2 (b). The potassium-doped C_{60} molecule reveals a superconducting transition at a temperature of 18 K. Some other, smaller fullerenes, such as C_{20} and C_{22}, and larger fullerene-like C_{70}, C_{72}, C_{76}, C_{80}, and C_{84} have also been found (Poole and Owens 2003). Fullerene dimers have been synthesized using high-speed vibration milling in the presence of potassium cyanide and reported by Komatsu et al. (Wang et al. 1997, Komatsu et al. 1998). A fullerene dimer is depicted in Figure 2.3.

2.2.2 Carbon Nanotubes

Carbon nanotubes are fascinating carbon nanostructures with numerous potential applications. The carbon nanotubes are formed by rolling the graphite sheet on the parallel axis of C–C bonds. The bonds available at the end of the graphite sheet close the carbon nanotube. Carbon nanotubes contain numerous structures with a variety of properties, with potential applications. Single-walled carbon nanotubes (SWCNTs) approximately 2 nm in diameter and 100 μm in length are known as nanowires. Even though carbon nanotubes are not made by rolling graphite sheets, it is the most promising way to describe the various structures for consideration in that the graphite sheets might be rolled into tubes along different axes to achieve multiple structures. There are four vectors illustrated in Figure 2.4 with the geometry of the graphite sheet: two basis vectors B_1 and B_2 of the 2-dimensional unit cell, one axis vector T and a circumferential vector C_h situated at right angles to vector T.

There are three different approaches to rolling up the graphite sheet about the axis of the T vector for producing carbon nanotubes. The first approach, in which T is parallel to the C–C bonds of the carbon hexagons (the *armchair* structure

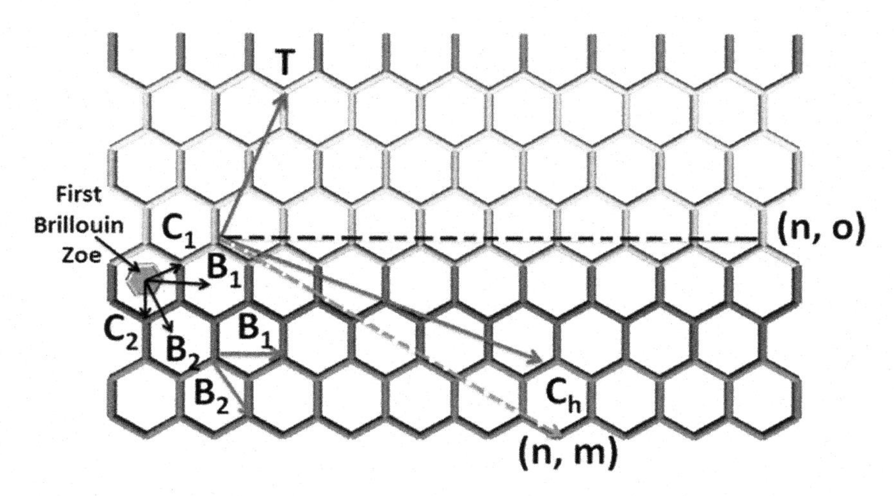

FIGURE 2.4 The orientation of a single layer of graphite plane with the basis vectors B_1 and B_2 of the two-dimentional unit cell, the axis vector T, and the circumferential vector C_h at right angles to T (by Nanotube Modular Software).

of carbon nanotubes) is generated as shown in Figure 2.5 (a). In the other two approaches, the axis of the T vector is not parallel to the C–C bonds. The carbon nanotubes are shown in Figures 2.5 (b) and (c), and described as the *zigzag* and the *chiral* structures, respectively, which are designed by rolling about a T vector involved in different directions of the graphite plane. The chiral structure retains a spiralling row of carbon atoms, as shown in Figure 2.5 (c). A pentagon carbon is usually available at each end of carbon nanotubes to close them. The complete tube is in the form of a cylinder, including half of the C_{60} fullerene configuration. Metal particles are generally used as catalysts in the formation of SWCNTs to control the successive growth of tubes, mostly found at the end of tubes (Mateo-Alonso 2006).

On the other hand, multi-walled carbon nanotubes (MWCNTs) exhibit many attention-grabbing features compared with SWCNTs, with enhanced electrical, chemical, and mechanical properties. The successive growth of MWCNTs is more accessible than the synthesis of SWCNTs, because MWCNTs can be grown without catalytic magnetic metal particles, which could change their magnetic and transport properties. The quantum interference phenomena, such as the Aharonov–Bohm effect, can be studied with larger-diameter MWNTs in magnetic fields. In contrast, a 600T magnetic field would be required to investigate the same phenomenon for SWNTs. The Russian-doll structure provides a prominent architecture for greater rigidity and mechanical stability in CNTs, illustrated in Figure 2.6, with doubled and multi-walled carbon nanotubes. This structure helps fabricate the tip of scanning probe microscopy. Furthermore, the configuration of MWCNTs is more suitable for synthesizing nanotube composites with chemically functionalized MWCNT walls (Ferro 2001).

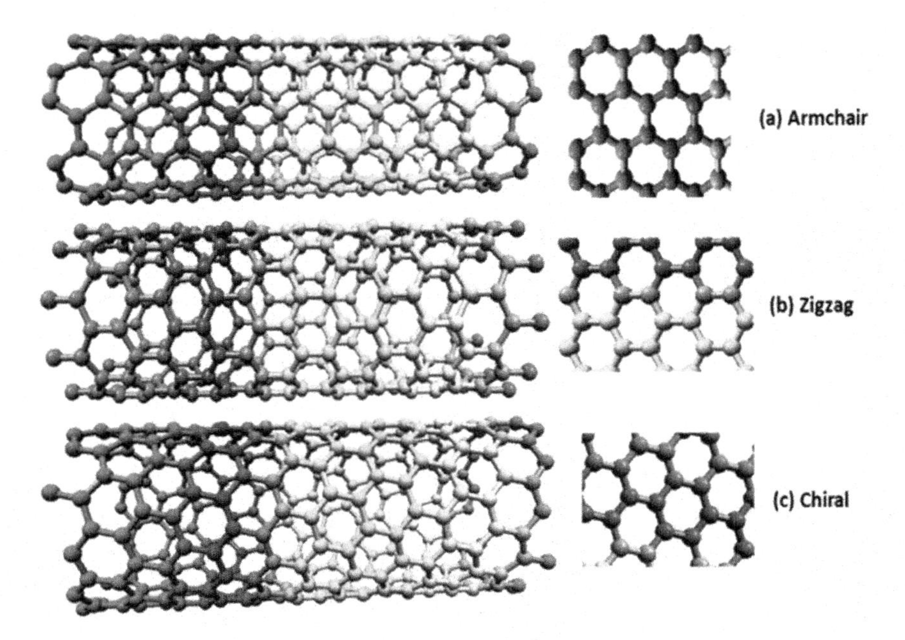

FIGURE 2.5 The structure of carbon nanotubes, depending upon how graphite sheets are rolled in (a) armchair, (b) zigzag, and (c) chiral structures (by Nanotube Modular Software).

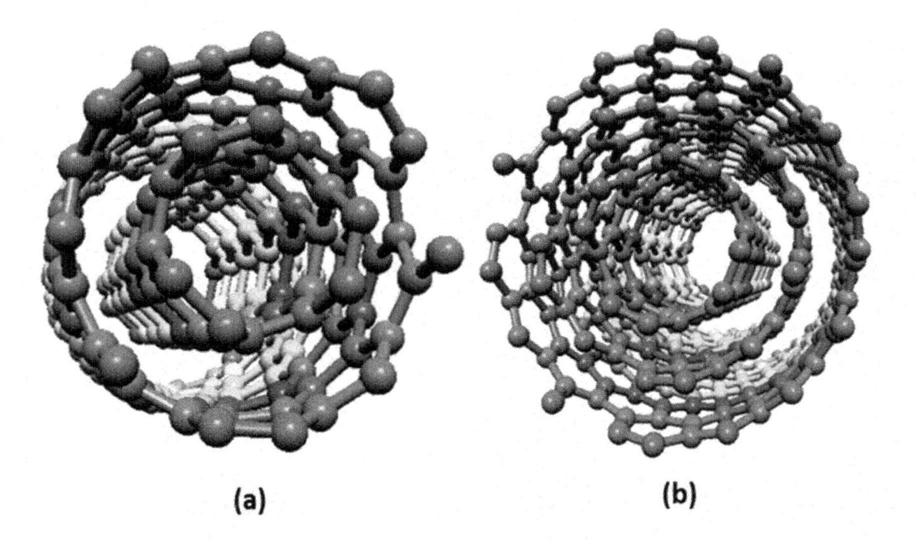

FIGURE 2.6 The schematic illustration of (a) double-walled and (b) multi-walled carbon nanotubes (by Nanotube Modular Software).

2.2.3 Graphene

Graphene is a single layer of graphite sheet or a 2D sheet of a conjugated carbon atom array in a honeycomb-shaped lattice. It is a versatile carbon nanomaterial with fascinating properties and potential applications. It is also known as a non-band gap semiconductor, in which the Fermi level can be tuned incessantly from the valence band to the conduction band. Every carbon atom contains four valence electrons. Three of these occupy the planar sp^2 hybrid orbital to form covalent in-plane (¾)-bonds. The remaining electron occupies a p_z orbital (the z-axis is perpendicular to the plane). Overlap between p_z orbitals results in free electrons, which can contribute to the electrical conductivity. The Bravais lattice of graphene with primitive vectors B_1 and B_2 is shown in Figure 2.4 (Oostinga 2010).

2.3 SECOND-GENERATION CARBON NANOMATERIALS

The rapid development of functionalized and doped carbon-based nanomaterials holds great technological promise for a range of applications. Polymer nanocomposites with nanofillers like CNTs, graphene, etc., have impressed researchers because of their unique and enhanced properties. Nanocarbons have also been embedded in the ferromagnetic and ferroelectric matrix for the enhancement of the efficiency of a variety of sensors. The functionalization and doping of nanocarbons have emerged as a fashion in nanotechnology of nanocomposites.

2.3.1 Functional Nanotube Polymer Nanocomposites

The functionalization of CNTs with different polymers has covered an enormous range of properties and applications. CNTs can enhance the electrical, mechanical, and thermal properties of a polymer matrix once these are introduced as a filler. The electrons and phonons can conduct without scattering along the 1D CNT structure, resulting in improved electrical and thermal conductivity, respectively, at room temperature. Several techniques have been developed for fabricating CNT/polymer nanocomposites, such as solvent casting, drop-casting, sonication, and melt mixing. In the solvent casting method, a suspension of nanotubes is prepared in a polymer and then the solvent is evaporated to create nanocomposites. A method was developed to achieve a MWNT/polystyrene (PS)/toluene suspension, followed by sonication, and casting it onto a dish to generate the nanocomposites, resulting in enhanced elastic modulus and breaking stress (Qian et al. 2000). Drop-casting was reported to enhance the electrical conductivity of SWNT/poly(methyl methacrylate) (PMMA)/toluene nanocomposites (Benoit et al. 2001). But there is a drawback to the solution casting technique: during the evaporation of the solvent, agglomeration of CNTs occurs, which disperse non-homogeneously in the polymer network (Kashiwagi et al. 2002). Bhattacharyya (2016) also faced the problem of agglomeration and non-homogeneity during the formation of SWNT/polypropylene (PP) nanocomposites by melt mixing. To solve this problem, Haggenmueller et al. (2000) first introduced the combination of solution casting and melt mixing methods to achieve improvements in the dispersion of SWNTs in the PMMA matrix.

Du et al. (2003) developed a coagulation method to achieve a homogeneous distribution of SWNTs in the PMMA matrix and produced better results (. CNT/ultrahigh molecular weight polyethylene (UHPE) nanocomposites were synthesized using wet mixing under ambient conditions, and achieved UHPE/CNT nanocomposites with a 20–50 times larger free zone surrounding CNTs on the external surface of PMMA (Al-Saleh 2015). A uniform coating layer of polyaniline (PANI)/MWCNT nanocomposites were designed by oxidative polymerization for ammonia detection at certain temperatures, with the sensitivity towards ammonia being found to be very high with a shorter recovery time, almost 6 seconds at 2 ppm (Abdulla et al. 2015). Zhang et al. (2016) and Yang et al. (2015) reported the preparation of polylactide (PLA)/MWCNT nanocomposites and achieved an excellent conductive network, with higher thermo-mechanical, heat distortion properties and an incremental increase in electrical conductivity of seven orders of magnitude, with considerable increases in Young's modulus.

2.3.2 Functional Graphene Polymer Nanocomposites

In the past two decades, the functionalized and doped graphene with polymers has also emerged in the form of versatile nanocomposites with an enormous range of properties and potential applications. There are numerous polymers for which their properties can be strengthened by introducing graphene as the filler. Eda and Chhowalla (2009) synthesized polystyrene (PS)/graphene nanocomposites using the solution blending method. PS acted as the host, and graphene worked as the filler molecule and achieved the highest mobility values after fabrication. Two separate processes were used by Bai et al. (2017) to introduce PS/graphene nanocomposites. The first process was the solution compounding method. The second process was a combination of solution compounding and the subsequent melt compounding method. They obtained low volume resistivity, high glass transition and improved thermal conductivity, with poor dispersion of PS in the nanocomposites prepared by the first process.

The 116% and 96% incremental increases in tensile strength and Young's modulus, respectively, compared to the pure epoxy with 1.00 wt% filling of graphene, were reported by Zhao et al. (2016) of the prepared epoxy composite loaded with epoxy/graphene. The percolation threshold was achieved at only 0.33 wt% loading, improving electrical conductivity from approximately 1^{-17} to 1^{-2} S cm^{-1} at about 2.00 vol% loading. Moreover, thermal conductivity was enhanced by 189% at 10 wt% epoxy/graphene filling. The well-dispersed graphene in the epoxy matrix in the form of epoxy/graphene nanocomposites was fabricated by Wajid et al. (2013), using freeze-dry/mixing and solution processing methods. The prepared composites exhibited improved electrical and mechanical properties, including enhanced thermal stability. The water solution processing was used to develop poly(vinyl alcohol) PVA/GO (graphene oxide) nanocomposites by Liang et al. (2009). The effective load transfers between the PVA matrix and nanofiller graphene were obtained with improved mechanical properties and fine dispersion of filler at the molecular level. Significantly, the insertions of only 0.7 wt% of GO achieved increases in Young's modulus up to 62% and in tensile strength up to 76%. The graphene and poly (vinyl alcohol) (PVA) were used in aqueous solution to prepare PVA/graphene nanocomposites (Zhao et al. 2010). They achieved an essential improvement in mechanical properties with the help of a significantly lower wt% of graphene and a 150% improvement in tensile strength by loading 1.8 vol% graphene in the PVA/graphene nanocomposites.

2.3.3 Functionalized Graphene and CNT With Other Nanoparticles

Nowadays, graphene and CNTs are functionalized with ferroelectric, ferromagnetic, and other fashionable materials to enhance the efficiency of devices for versatile applications. Xia et al. (2012) fabricated $CoFe_2O_4$/graphene nanocomposites with different graphene contents by hydrothermal processes. They found that the nanocomposites prepared exhibited significantly altered electrochemical performance as anode material for lithium-ion batteries. The presence of 20 wt% graphene in the $CoFe_2O_4$/graphene nanocomposites electrode can transport a high reversible specific capacity up to 1082 mAh g^{-1}, including excellent rate capability and cycling stability.

A novel piezoelectric nanogenerator (PNG) was fabricated by Yaqoob et al. (2017), using poly (vinylidene fluoride) (PVDF), barium titanate (BTO), and surface-modified *n*-type graphene (*n*-G). The PVDF/BTO/*n*-G tri-layer PNG exhibits a maximum output voltage of 10 V at an applied force of 2 N and a current of 2.5 μA and a power of 5.8 μW at 1 MΩ load resistance. Furthermore, the fabricated device was able to demonstrate good stability for future piezoelectric generating technologies. Huang et al. (2017) synthesized $CoFe_2O_4$ (CFO) nanoparticles (NPs) onto nitrogen-doped activated carbon (N-AC), using a hydrothermal-assisted annealing assembly technique. They reported that the CFO/N-AC nanocomposites prepared possessed rapid electron transport with less resistance and excellent electro-catalytical activity for the oxygen reduction reaction, due to the large surface area. They achieved maximum power density ~1770.8 ± 15.0 mW m^{-2}, which is 2.39 times higher than the control AC. Hull et al. (2006) functionalized MWCNTs with –C=O, –C–O–C, –COO– and –C–OH groups to create new catalyst materials for fuel cell applications. They used a sonochemical treatment process to deposit Pt nanoparticles of 4 nm diameter under an aqueous acidic solution environment and form Pt-MWCNT interfaces as the catalyst.

2.4 THIRD-GENERATION NANOCARBON MATERIALS

Combining different allotropic forms of carbon at the nanoscale fabricates modified surfaces with exclusive properties. Polymer–nanocarbon, nanocarbon–nonocarbon and nanoparticle–nanocarbon hybrid nanomaterials are discussed in detail in Subsection 2.4, with their enhanced properties and promising applications.

2.4.1 Polymer–Nanocarbon Hybrid Nanomaterials

Carbon aerogel is an excellent example of the hybridization of carbon nanoparticles. In 1989, Pekala developed the first polymer carbon aerogels (PCAs) (Pekala 1989). He demonstrated that, under alkaline conditions with formaldehyde, the polycondensation of resorcinol generates the formation of surface hybridized clusters of polymer. The covalent cross-linking of these clusters yields gels treated under supercritical conditions and generates organic aerogels of $\leqslant 0.1$ g cm^{-3} density. The aerogels contain interrelated colloidal particles of ~10 nm diameter, which are transparent and dark red in colour. Wang et al. (2010) developed a hybrid material of graphene/polyaniline using polymerization-reduction/dedoping-redoping processes to fabricate electrodes for a supercapacitor. Compared with pure, individual components, they attained a high specific capacitance of 1126 F g^{-1}, which was achieved with a retention life of 84% after only 1,000 cycles for supercapacitors, with better electrochemical performances and improvements in the power density and energy density.

2.4.2 Nanocarbon-nanocarbon Hybrid Nanomaterials

Over the diamond-like carbon (DLC) substrate, MWCNTs and carbon nanofibers were also deposited to make carbon-based nanohybrid materials to enhance the geometry and chemistry of the nanohybrid matrix. This nanohybrid is very useful in detecting specific biomolecules, such as glutamate, ascorbic acid, and dopamine, significant neurotransmitters in the mammalian central nervous system (Laurila et al. 2017). For fused filament fabrication (FFF), acrylonitrile butadiene styrene (ABS) hybrid nanocomposite filaments were integrated with graphene nanoplatelets (GNPs) and CNTs by Dul et al. (2020). They produced 6 wt% of total GNPs and/or CNT nanofiller and relative percentage ratios of 0, 10, 30, 50, 70, or 100% GNP/CNT, using melt mixing and compression moulding mechanics. The hybrid nanocomposites with GNP content revealed a linear increment in modulus and a decrement in strength. In contrast, the addition of CNTs to hybrid nanocomposites determined an optimistic increase in electrical conductivity with possibly perilous decreases in melt flow index. Moreover, because of the promising improvement in mechanical and electrical properties, 50:50 GNP/CNT nanohybrid composition was designated as the most appropriate content for the filament manufacturing of 6 wt% carbonaceous nanohybrids.

2.4.3 Other Nanoparticle-nanocarbon Hybrid Nanomaterials

To develop a gold nanoparticle (GNP)/MWCNT nanohybrid, Zhang et al. (2006) demonstrated a new approach, in which they attached sodium dodecyl sulfate (SDS) as a surfactant in successive layers with MWCNTs on an indium tin oxide (ITO) substrate and then attached GNPs on the surface of MWCNTs. A crucial step was taken by Lee et al. (2014) for generating an ordered multileveled hybrid nanocomposite, using CNTs ($d \approx 1.2$ nm)/Ag nanowire (AgNW) ($d \approx 150$ nm) as highly flexible, stretchable, and transparent conductors with improved optical, electrical, and mechanical properties.

They suggested that these highly elastic CNT/AgNW, flexible, transparent nanohybrid conductors can be planted on any soft non-planar surfaces to achieve an electronics interface for future fashionable electronics. Zhang et al. (2016) fabricated a tin dioxide (SnO_2)/reduced graphene oxide (RGO) hybrid nanocomposite humidity sensor on a polyimide flexible substrate through a facile one-step hydrothermal route. They compared this novel sensor with traditional humidity sensors and demonstrated that the SnO_2 modified graphene sensor showed considerable sensitivity and a quick response/recovery time over a complete humidity range measurement.

2.5 REFERENCES

Abdulla, S., Mathew, T.L., Pullithadathil, B. 2015. Highly sensitive, room temperature gas sensor based on polyaniline-multiwalled carbon nanotubes (PANI/MWCNTs) nanocomposite for trace-level ammonia detection. *Sensors and Actuators B Chemicals* 221:1523–1534.

Al-Saleh, M.H. 2015. Influence of conductive network structure on the EMI shielding and electrical percolation of carbon nanotube/polymer nanocomposites. *Synthetic Metals* 205:78–84.

Bai, Q.-Q., Wei, X., Yang, J.-H. et al. 2017. Dispersion and network formation of graphene platelets in polystyrene composites and the resultant conductive properties. *Composites Part A: Applied Science and Manufacturing* 96:89–98.

Benoit, J.M., Corraze, B., Lefrant, S., Blau, W., Bernier, P. and Chauvet, O. 2001. Transport properties of PMMA-carbon nanotubes composites. *Synthetic Metals* 121:1215–1216.

Bhattacharya, M., 2016. Polymer nanocomposites—a comparison between carbon nanotubes, graphene, and clay as nanofillers. *Materials 9*(4): 262.

Du, F., Fischer, J.E. and Winey, K.I. 2003. Coagulation method for preparing single-walled carbon nanotube/poly(methyl methacrylate) composites and their modulus, electrical conductivity, and thermal stability. *Journal of Polymer Science: Part B: Polymer Physics* 41:3333–3338.

Dul, S., Ecco, L.G., Pegoretti, A. and Fambri, L. 2020. Graphene/carbon nanotube hybrid nanocomposites: effect of compression molding and fused filament fabrication on properties. *Polymers* 12:101.

Eda, G., Chhowalla, M. 2009. Graphene-based composite thin films for electronics. *Nano Letters* 9:814–818.

Ferro, L. 2001. Physical properties of multiwalled nanotubes. In *Carbon Nanotubes*, ed. Ph. Avouris, 329–391. Topics in Applied Physics. Berlin Heidelberg: Springer.

Haggenmueller, A., Gommans, H.H., Rinzler, A.G., Fischer, J.E. and Winey, K.I. 2000. Aligned single-wall carbon nanotubes in composites by melt processing methods. *Chemcal Physics Letters* 330:219–225.

Huang, Q., Zhou, P., Yang, H., Zhu L. and Wu, H. 2017. In situ generation of inverse spinel $CoFe_2O_4$ nanoparticles onto nitrogen-doped activated carbon for an effective cathode electrocatalyst of microbial fuel cells. *Chemical Engineering Journal* 325:466–473.

Hull, R.V., Li, L., Xing, Y. and Chusuei, C.C. 2006. Pt nanoparticle binding on functionalized multiwalled carbon nanotubes. *Chemical Materials* 18:1780–1788.

Inagaki, M. 2000. *New Carbons Control of Structure and Functions*. Toyota: Elsevier Science Ltd.

Kashiwagi, T., Grulke, E., Hilding, J., Harris, R., Awad, W. and Douglas, J. 2002. Thermal degradation and flammability properties of poly(propylene)/carbon nanotube composites. *Macromolecular Rapid Communications* 23:761–765.

Komatsu, K. et al. 1998. Mechanochemical synthesis and characterization of the fullerene dimer C_{120}. *Journal of Organic Chemistry* 63:9358–9366.
Laurila, T., Sainio, S. and Caro, M.A. 2017. Hybrid carbon based nanomaterials for electrochemical detection of biomolecules. *Progress in Materials Science* 88:499–594.
Lee, P., Ham, J., Lee, J. et al. 2014. Highly stretchable or transparent conductor fabrication by a hierarchical multiscale hybrid nanocomposite. *Applied Functional Materials* 24:5671–5678.
Liang, J., Huang, Y., Zhang, L. et al. 2009. Molecular-level dispersion of graphene into poly (vinyl alcohol) and effective reinforcement of their nanocomposites. *Advanced Functional Materials* 19:2297–2302.
Mateo-Alonso A. 2006. Fullerenes and their derivatives. In *Carbon Nanomaterials*, ed. Y. Gogotsi, 3–33. Pennsylvania, PA: CRC Press/Taylor & Francis.
Oostinga, J.B. 2010. *Quantum Transport in Graphene.* Delft-Leiden: Gildeprint Drukkerijen.
Pekala, R.W. 1989. Low density, resorcinol-formaldehyde aerogels, US Patent 4873218.
Pekala, R.W. 1989. Organic aerogels from the polycondensation of resorcinol with formaldehyde, *Journal of Materials Science* 24:3221–3227.
Poole Jr, C.P. and Owens, F.J. 2003. *Introduction to Nanotechnology.* Hoboken, NJ: Wiley.
Qian, D., Dickey, E.C., Andrews, R. and Rantell, T. 2000. Load transfer and deformation mechanisms in carbon nanotube-polystyrene composites. *Applied Physics Letter* 76: 2868–2870.
Wajid, A.S., Ahmed, H.S.T., Das, S. et al. 2013. High-performance pristine graphene/epoxy composites with enhanced mechanical and electrical properties. *Macromolecular Material Engineering* 298:339–347.
Wang, G.-W. et al. 1997. Synthesis and X-ray structure of dumb-bell-shaped C_{120}. *Nature* 387: 583–586.
Wang, H., Hao, Q., Yang, X. et al. 2010. A nanostructured graphene/ polyaniline hybrid material for supercapacitors. *Nanoscale* 2:2164–2170.
Xia, H., Zhu, D. and Fu, Y. 2012. $CoFe_2O_4$-graphene nanocomposite as a high-capacity anode material for lithium-ion batteries. *Electrochimica Acta* 83:166–174.
Yang, J.H., Lee, J.Y. and Chin, I-J. 2015. Reinforcing effects of poly (D-Lactide)-g-multiwall carbon nanotubes on polylactide nanocomposites. *Journal of Nanoscience and Nanotechnology* 15:8086–8092.
Yaqoob, U., Iftekhar Uddin, A.S.M. and Chung, G-S. 2017. A novel tri-layer flexible piezoelectric nanogenerator based on surface- modified graphene and PVDF-$BaTiO_3$ nanocomposites. *Applied Surface Science* 405:420–426.
Zhang, D., Chang, H., Li, P., Liu, R. and Xue, Q. 2016. Fabrication and characterization of an ultrasensitive humidity sensor based on metal oxide/graphene hybrid nanocomposites. *Sensors and Actuators B: Chemical* 225:233–240.
Zhang, K., Peng, J.-K., Shi, Y.-D. et al. 2016. Control of the crystalline morphology of poly (Llactide) by addition of high-melting-point poly(L-lactide) and its effect on the distribution of multiwalled carbon nanotubes. *Journal of Physics and Chemistry* B120:7423–7437.
Zhang, M. Su, L. and Mao, L. 2006. Surfactant functionalization of carbon nanotubes (CNTs) for layer-by-layer assembling of CNT multi-layer films and fabrication of gold nanoparticle/CNT nanohybrid, *Carbon* 44:276–283.
Zhao, S., Chang, H., Chen, S. et al. 2016. High-performance and multifunctional epoxy composites filled with epoxide-functionalized graphene. *European Polymer Journal* 84:300–312.
Zhao, X., Zhang, Q., Chen, D. et al. 2010. Enhanced mechanical properties of graphene-based poly (vinyl alcohol) composites. *Macromolecules* 43:2357–2363.

3 Carbon Dots

Structure, Synthesis, Properties, and Advanced Health Care Applications

Chinnu Sabu, V. K. Ameena Shirin, Renu Sankar, and K. Pramod

CONTENTS

DOI: 10.1201/9781003110781-3

3.1 INTRODUCTION

Carbon dots (CDs) are a novel class of spherical small-sized nanomaterials with diameters of less than 10 nm. Because of their attractive properties, they hold a significant position in the current research field. The unique properties of CDs include optical properties, economical cost of production, biocompatibility, and ease of synthesis (Bhartiya et al. 2016). The CDs were discovered during the purification of single-walled carbon nanotubes (SWCNTs); it was later discovered that surface passivation of these nanoparticles would enhance their fluorescent emission (Sun et al. 2006). Subsequently, much research has explored the structure, properties, and applications of CDs.

CDs can be synthesized from either natural or synthetic sources. Their most important novel characteristic is the ability to be synthesized from lower-cost sources like small organic molecules. Synthesis from readily available and cheap sources has further enhanced their cost-effectiveness and sustainability (de Medeiros et al. 2019). Currently, there are many methods for synthesising CDs, categorized as either top-down or bottom-up approaches. The top-down process refers to the breaking down of large carbon structures into smaller particles. In contrast, the bottom-up process refers to the synthesis of CDs from small particles or organic molecules (Zhang and Yu 2016).

CDs have a wide range of applications in bioimaging, biosensing, drug delivery, catalysis, and other healthcare fields. CDs' fluorescent nature and biocompatibility enable their use in both *in vitro* and *in vivo* bioimaging and for several diagnosis purposes (Liu et al. 2012). The doping of CDs with atoms like nitrogen and sulphur were carried out to enhance their application. Due to their tunable physical and chemical characteristics, they have many biomedical applications. Their small size, biocompatibility, and versatile nature enable them to be potential carriers for drug delivery (Sun et al. 2006). Their stability in aqueous media and their ability to undergo surface modification further extends their biological applications (Lim et al. 2015). Moreover, they are also used to detect several ions like Cu^{2+}, Hg^{2+}, NO_3^-, Fe^{2+}, etc. (Sharma and Das 2019).

This chapter explains in detail the structure, properties, and advanced healthcare applications of CDs.

3.2 STRUCTURE

CDs are an emerging class of nanomaterial, with a diameter of less than 10 nm. They are mostly spherical in nature, and some are hollow structured. Several studies have been conducted to gain knowledge about the structure of CDs. They are sp^2 hybridized, and some are sp^3 hybridized also. Studies also confirm the existence of CDs with diamond-like structures (Bhartiya et al. 2016). The structure mainly depends on the method adopted for the synthesis of the CDs. They show a mainly spherical morphology and sometimes also a crystalline or amorphous one (Kelarakis 2014). They can be surface functionalized with many functional groups like –OH, C=O, $-NH_2$, etc., to bind with many organic and inorganic molecules. Also, both hydrophilic and hydrophobic molecules can be used for surface functionalization. Functionalization helps to achieve more properties and applications (Gayen et al. 2019).

CDs mainly consist of only C,H and O. Their compositions are analyzed commonly by using Fourier transform infrared spectroscopy (FTIR) or X-ray photoelectron spectroscopy (XPS) spectroscopy. CDs composition consists of both interlayer spacing and in-plane lattice spacing. The crystalline nature of CDs can be analyzed by using Powder X-Ray diffraction (PXRD), Raman spectroscopy, etc. (Kang et al. 2019).

3.3 SYNTHESIS

There are several methods available for the synthesis of CDs. These methods are mainly classified into two types: (1) top-down and (2) bottom-up. Figure 3.1 presents the different methods of synthesis. Table 3.1 summarizes the different methods of synthesis

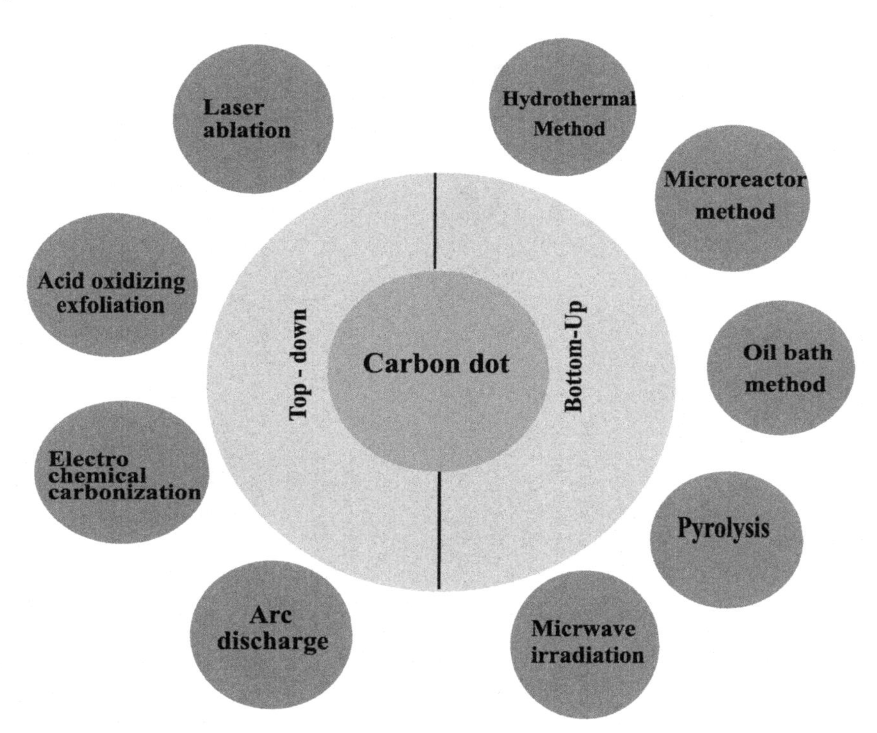

FIGURE 3.1 Different methods of synthesis of CDs.

TABLE 3.1
Different Methods of Synthesis

Synthesis Method	Source	Quantum Yield	Merits	Demerits	Ref.
Laser ablation	Graphite powders	4% to 10%	Fast, highly tunable, very effective	Modification is necessary. Low quantum yield, and very little control over the size	(Li et al. 2011; Nguyen et al. 2015)
Electrochemical carbonization	Carbon nanotube, sodium citrate, and urea	25.6%	One-step method, size is controllable and stable.	Few precursors.	(Ming et al. 2012; Zhou et al. 2007; Liu et al. 2016)
Arc discharge	Carbon nanotube, soot	1.6%	Good water solubility, larger particle size distribution	Due to larger particle size, they decrease the specific surface area of CDs, which limits the active reaction sites of CDs	(Wang et al. 2019; Roy et al. 2015)
Acid-oxidizing exfoliation	Carbon soot, Chinese ink	60–80 %	High quantum yield, tunable photoluminescent properties, and longer lifetime	Hydrothermal treatment required	(Yang et al. 2014)
Microwave irradiation	Citric acid, glycerol	11.7%	Inexpensive, eco-friendly, and fast	Poor control over size	(Liu et al. 2014; Wang et al. 2014)
Hydrothermal method	L-lysine and thiourea	53.19%	Non-toxic, inexpensive, and eco-friendly	Poor control over size	(Yang et al. 2011a; Khan et al. 2020)
Pyrolysis	Glycerol, sodium alginate	88%	Simple, economical, mass production of CDs	High temperature is required	(Ma et al. 2019; Fong et al. 2016)
Oil bath method	Citric acid	20.3%	Fast and convenient method		(Zhou et al. 2019)
Microreactor method	Citric acid	60%	Rapid and continuous method		(Berenguel-Alonso et al. 2019; Rao et al. 2017)

3.3.1 Top-down Synthesis

Top-down synthesis of CDs involves the breaking-up of larger carbon precursors like graphite and carbon nanotube into smaller carbon structures.

3.3.1.1 Laser Ablation

This is the first method used for the production of carbon dots in a controlled manner. Interaction between a pulsed beam and the solid precursor results in the ejection of nanoparticles (Sun et al. 2006). Here, the carbon material precursor absorbs high energy from the laser pulse, resulting in the stripping of electrons from atoms through thermionic and photoelectric emissions. Then, the breaking of carbon dots occurs due to the high repulsive force between the positive ions and the solid materials by the high electric force (Xiao et al. 2017). Changing laser ablation parameters can improve carbon dot characteristics. The size of the CDs and its photoluminescence properties can be controlled by changing parameters like irradiation time. The very small-sized CDs can be synthesized by increasing the irradiation time in the laser ablation technique (Nguyen et al. 2015). During the one-step synthesis of CDs from graphite flakes in a polymer solution, an increase in the pulse duration can change the CDs size from 3 nm to 13 nm. Varying pulse duration changes the nucleation conditions and thereby changes the size distribution. After this synthesis procedure, no passivation procedure is needed because surface passivation occurs simultaneously along with the formation of the CDs (Hu et al. 2011). Ablation of graphite with UV radiation of wavelength 355 nm can result in the formation of CDs without any aggregates. But lots of carbon aggregates are produced when using UV radiation of wavelength 532 nm (Reyes et al. 2016).

3.3.1.2 Arc Discharge

The arc discharge method for the synthesis of CDs originated in 2004. Three types of CDs with different fluorescent properties and molecular mass were accidentally obtained during the synthesis of single-walled carbon nanotubes (SWNTs) by the arc discharge method. These prepared CDs can emit orange, yellow, or blue-green fluorescence at 365 nm (Xu et al. 2004). This method is used to prepare CDs with a quantum yield of 1.6°C from crude carbon nanotube. Crude materials were subjected to oxidation with nitric acid. After oxidation, the oxidized materials were extracted with a pH 8.4 alkaline solution, and, finally, the extracted materials were purified by gel electrophoresis (Roy et al. 2015). This method reorganizes the carbon atoms which decompose from the carbon precursors at the anode. These carbon atoms from the precursors are driven by the gas plasma in the sealed reactor. High-energy plasma is generated due to the high temperature in the reactor. CDs are formed at the cathode by the assembly of carbon vapour (Arora and Sharma 2014). CDs obtained by this method have high water solubility and larger particle size distribution. Due to their larger particle size, they decrease the specific surface area of the CDs, which limits the active reaction sites of CDs (Wang et al. 2019).

3.3.1.3 Electrochemical Carbonization

The electrochemical method is the simplest and most convenient synthesis method. This method can be carried out at normal temperatures as well as under highly pressurized conditions. This method is widely used because of the easy regulation of the particle size of the CDs formed by this method (Ahirwar et al. 2017). Compared to the arc discharge method, this method is more effective and scalable because various carbonic materials can be used as the carbon precursor without using any strong acids and purification processes. It involves a redox reaction in an electrochemical cell under an electric current applied between two electrodes separated by an electrolyte. Graphitic CDs were first synthesized by this method; here, multiwalled carbon nanotubes (MWCNTs) were used as an electrode, resulting in the formation of graphitic CDs with blue luminescence (Zhou et al. 2007). Electrochemical carbonization of sodium citrate and urea in deionized water results in the synthesis of blue light-emitting CDs with a particle size of 2.4 nm (Hou et al. 2015). It is the most common route among top-down synthesis methods because of its remarkable advantages, such as high purity and yield, low cost, and easy tuning of the CDs size. CDs with a diameter of 4.2 nm were synthesized using graphite electrodes as a carbon source by this method (Liu et al. 2016).

3.3.1.4 Acid-oxidizing Exfoliation

Acid-oxidizing exfoliation is mainly used for the exfoliation and decomposition of bulk carbon into small nanoparticles, while also resulting in the introduction of hydrophilic groups (hydroxyl or carboxyl groups) on the surface of the CDs formed. Such introduction of hydrophilic groups can improve the water solubility of the CDs and also its fluorescent properties (Zhang et al. 2017). Heteroatom doped CDs have been synthesized by this method. Here, carbon nanoparticles were obtained from Chinese ink and then subjected to oxidation by the use of a solution of nitric acid, sulphuric acid, and sodium chlorate. Then, the oxidized materials are subjected to hydrothermal reduction with dimethylformamide (DMF), sodium hydrosulphide (NaHS), and sodium selenide (NaHSe) as the sources of nitrogen, sulphur, and selenium, respectively. The CDs obtained exhibit high quantum yield, tunable photoluminescent properties, and longer lifetime (Yang et al. 2014).

3.3.2 Bottom-up Synthesis

The bottom-up synthesis of CDs involves the synthesis of small carbon units into CDs of the required size by the application of some form of energy. This approach includes microwave irradiation, hydrothermal/solvothermal treatment, pyrolysis, oil bath method, or microreactor method for the synthesis of CDs.

3.3.2.1 Microwave Irradiation

The microwave irradiation method is the most rapid and cheapest method for the synthesis of CDs. Here, the mixture of precursor materials is exposed to irradiation with electromagnetic waves, which have a wavelength within the range 1 mm to 1 m

(Gong et al. 2014). Hydrophilic, hydrophobic, or amphiphilic particles can be synthesized by this method. This method has several advantages over other methods, of which the most significant is the reduced time requirement. Also, the uniform heating in the microreactor enables rapid reactions, reduced side effects, and a smaller number of by-products. In addition to this, there are some disadvantages, like limited applications in small-scale reactions and decreased usability with solvents having a low boiling point (de Medeiros et al. 2019). The fluorescence intensity of the CDs can be enhanced by increasing the pH of the solution. Recent work has demonstrated the synthesis of CDs from the ashes of the eggshell membrane by one-minute micro-irradiation in a sodium hydroxide solution. The CDs obtained had a diameter of 3.88 nm (Jusuf et al. 2018).

The CDs prepared from the one-step microwave irradiation of citric acid, using tryptophan as both a passivating agent and a source of nitrogen, showed enhanced water solubility (Wang et al. 2014). This method enabled the synthesis of CDs from banana peel within 5 min at 700 W. The CDs prepared were used for the quantification of glucose in blood plasma (Gul et al. 2020). Green-fluorescent CDs were prepared by irradiation for 1 min using sucrose as a source of carbon and diethylene glycol as the reaction medium (Jaiswal et al. 2012).

3.3.2.2 Hydrothermal Method/Solvothermal Treatment

This is the most commonly used method for the synthesis of CDs. The advantages of this method over others include low cost and environmentally friendly routes for the synthesis of CDs. This approach can be used for the synthesis of CDs from saccharin, amine, organic acids, and derivatives of organic acids (Yang et al. 2011a,b). Here, the substrate is heated to a temperature of 180°C for a period of 4 h–6 h in an autoclave and then purified by centrifugation or dialysis. Finally, the diameter of the obtained CDs is measured. Initially, the CDs were prepared by this method, using ascorbic acid as the small organic molecule, which was heated for 4 h. Then, the solution obtained was purified by dialysis and the synthesized CDs had a diameter of 2 nm (Zhang et al. 2010). Nitrogen- and sulphur-doped CDs (NS-CDs) were prepared by using amino acid as a source. L-serine and L-cystine were used as an amino acid source and CDs obtained had a diameter of nearly 2.6 nm (Zeng et al. 2015).

Using L-lysine and thiourea as precursors, NS-CDs were prepared by this strategy. The yield and diameter of the prepared CDs were found to be 53.19% and 6.86 nm, respectively. These synthetic CDs were shown to be effective in the detection of picric acid in aqueous solutions (Khan et al. 2020). Blue photoluminescent CDs prepared by the one-pot hydrothermal process, using L-histidine and citric acid as precursors, showed increased solubility in water and tolerance to water. The CDs prepared in this way helped in the quantitative determination of chlorogenic acid (Han et al. 2020).

3.3.2.3 Pyrolysis

Pyrolysis involves the irreversible thermal decomposition of the precursor materials in an inert environment under the conditions of high temperature and controlled pressure. Here, as a result of physicochemical changes, a residue containing carbon is produced (Sharma and Das 2019). This method is simple and economical, which helps in the mass production of CDs. The pyrolysis of glycerol results in the production of highly emissive CDs (Lai et al. 2012).

The thermal pyrolysis method was used for the synthesis of CDs from citric acid, using diethylenetriamine as a passivating agent, which results in the production of CDs within the size range of 5–8 nm (Feng et al. 2016). CDs doped with nitrogen have been developed by this method with outstanding stability and a high fluorescence quantum yield of nearly 88%. The CDs prepared have excellent optical properties even without the purification process (Ma et al. 2019).

Fennel seeds are an important source of carbohydrate which can be used for the synthesis of CDs by the process of pyrolysis. Here, the fennel seeds were crushed and the powder was transferred to a crucible, then heated at a temperature of 500°C for about 3 h and cooled to room temperature. The product obtained was dissolved in deionized water and sonicated for 5 min. Finally, the solution produced was centrifuged, filtered, and purified by dialysis. The CDs produced had high colloidal solubility, photo-stability, and environmental stability. Such CDs find applications in bio-sensing, cellular imaging, sensing, etc (Dager et al. 2019).

3.3.2.4 Oil Bath Method

The microwave-assisted oil bath method is a fast and convenient way of achieving the synthesis of CDs. Nitrogen-doped CDs were synthesized by using citric acid as a source of carbon and ethylenediamine as a nitrogen source. The quantum yield was found to be 20.3% and doping with nitrogen and sulphur increased the quantum yield more. In this method, the components were mixed with ultra-pure water in a conical flask that was connected to a condenser. Then, the flask was placed in a microwave oven and then cooled to room temperature. Subsequently, the mixture was placed in an oil bath at 200^{0}C for 10 min and finally purified by dialysis (Zhou et al. 2019).

3.3.2.5 Microreactor Method

This method allows the rapid and continuous synthesis of CDs on a large scale. In a study, CDs were prepared by using different microreactors by maintaining optimal conditions of concentration, reaction temperature, and flow rate (i.e., 1.5 ml, 160°C, and 16 ml/min,). It takes less than 5 min to produce CDs with a percentage yield of 60%. Further investigations proved that the CDs prepared could be used as nanoprobes for cell imaging and Fe^{3+} detection (Rao et al. 2017).

Low-Temperature Co-fired Ceramic (LTCC) microreactors can be used for the efficient synthesis of CDs. Such a microreactor consists of an optical window and a heater. The optical window enables the continuous monitoring of reactions taking place inside the reactor. The window has the capacity to withstand even extreme reaction conditions of 190°C and 17 bar pressure. This method can be used for the mass synthesis of different types of CDs by the hydrothermal method. Here, citric acid and different nitrogen sources were used as precursors. Due to the low toxicity of the CDs produced, it finds applicability as a contrasting agent in bioimaging (Berenguel-Alonso et al. 2019).

3.4 PROPERTIES OF CDs

3.4.1 Physicochemical Properties

CDs are zero-dimensional photoluminescent carbon nanoparticles with a diameter of less than 10 nm (Liu et al. 2011). Transmission electron microscopy and atomic force microscopy can be used to characterize the morphology of CDs. The crystalline characteristics of CDs can be determined by X-ray diffraction patterns, Raman spectroscopy, and high-resolution transmission electron microscopy. The X-ray diffraction patterns show a broad peak depicting the highly disordered carbon atoms (Wu et al. 2017). Two broad peaks are observed for the Raman spectrum of CDs at 1300 cm^{-1} and 1580 cm^{-1} caused by the disordered D band and the crystalline G band due to sp3 defects and in-plane vibration of sp2 carbon atoms, respectively. The relative intensity of the characteristic band can be used to determine the degree of graphitization and crystallization of the interior of the CDs (Qu et al. 2013).

CDs are composed of elements like C, H, and O due to the presence of carboxylic acid moieties, resulting in improved water solubility. The presence of a carboxylic acid group offers an opportunity for functionalization with different groups. Intercalation of CDs with heteroatoms helps to tune the conduction band position and offers unexpected functions (Wu et al. 2017).

3.4.2 Optical Properties

CDs have become an efficient short-wavelength photon-harvester due to the absorption of π electrons in the sp^2 framework (Yuan et al. 2016). Strong optical absorption is observed in the UV region, with strong peaks at 230 and 300 nm due to aromatic C=C bonds and C=O bonds, respectively (Zhao et al. 2015; Anwar et al. 2019). Moreover, surface modification or functionalization of CDs results in varying optical properties (Wang et al. 2019). In addition, the variation in size, structure, or composition of hybridization derivatives can alter the absorption characteristics (Yuan et al. 2016).

One of the fascinating properties that play a crucial role in experimental and practical applications is the tunable photoluminescence characteristics of CDs. The peak of photoluminescence spectra shows excitation-dependent properties (Ge et al. 2014; Tang et al. 2014). Furthermore, the quantum yield of CDs depends on the fabrication method and the surface chemistry. The quantum yield is low for unpassivated CDs, as the electron-withdrawing groups of carboxylic and epoxy groups can reduce the π electron cloud density of CDs. However, surface functionalization with electron-donating groups can improve the π electron cloud density of CDs and further increase the quantum yield (Yuan et al. 2016). In addition, doping of CDs with heteroatoms can effectively modulate bandgap and electron density (Li et al. 2012; Shuhua Li et al. 2014; Yuan et al. 2015). The chemiluminescence intensity is directly proportional to the concentration of CDs with a certain absorption range (Mishra et al. 2018).

3.4.3 Electronic Properties

Quantum confinement effects and edge effects play a pivotal role in excellent electronic characteristics for CDs. It is observed that photoinduced CDs can act as electron donors as well as electron acceptors (Yuan et al. 2016). These photoinduced electron transfer properties of CDs offer a new pathway in light energy conversion applications. The electronic energy relaxation pathways were first studied by Li et al. in CDs of uniform size and observed that the photoexcited CDs had a strong possibility of relaxation in triplet states and emitted both fluorescence and phosphorescence with relative amplitude at room temperature, based on the energy of excitation. Because of the considerably long lifespan and reactivity of electronic triplet states, the outputs may have major consequences for CDs uses (Mueller et al. 2010).

3.4.4 Biological Properties

CDs are found to be biocompatible in different forms, with little or no cytotoxicity, as the core of CDS contain only non-toxic carbon atoms which are crucial for various bio-applications (Jaleel and Pramod 2018). The immune reaction can be triggered by a greater concentration of CDs, which further raises the rate of the cluster determinant factor and interferons. However, these stimulations do not induce any sort of morphological alterations in immune organs (Yang et al. 2009). Various findings of acute, subacute, and genotoxicity experiments indicate that the health of laboratory animals is not negatively impacted by CDs at various concentrations. Moreover, there was hardly any major variation in the biological markers (Wang et al. 2013). In addition, cytotoxicity tests have demonstrated no notable effects on the cell viability of cancer cell lines (Zhang and Yu 2016). In addition, the *in-vivo* kinetic study behaviour of CDs on mice by the intravenous route showed negligible toxicity and high biocompatibility (Huang et al. 2013).

3.5 ADVANCED HEALTH CARE APPLICATIONS OF CDS

3.5.1 Antimicrobial Agents

CDs have been found to be non-toxic both *in vivo* and *in vitro*. CDs as such do not have antibacterial activity, but functionalization with different agents can help CDs exhibit antibacterial properties. The photoexcited CDs can act as antimicrobial agents by generating reactive oxygen species (ROS). The different mechanisms by which CDs elicit antimicrobial effects include the attachment of CDs to the surface of bacteria, the photoinduced generation of ROS, the breakage of the bacterial cell wall, DNA/RNA damage due to ROS-induced oxidative stress resulting in inhibition of the expression of various genes, and the induction of oxidative damage to proteins (Li et al. 2016; Al Awak et al. 2017; Hao Li et al. 2018). Under visible/natural light, CDs can generate ROS, which can destroy cell components and ultimately lead to cell death (Ipe et al. 2005; Jhonsi et al. 2018).

2,2-(Ethylenedioxy)bis(ethylamine)(EDA) CDs under visible/ UV light inhibit the growth of *Escherichia coli* (Al Awak et al. 2017). CDs prepared by an electrochemical

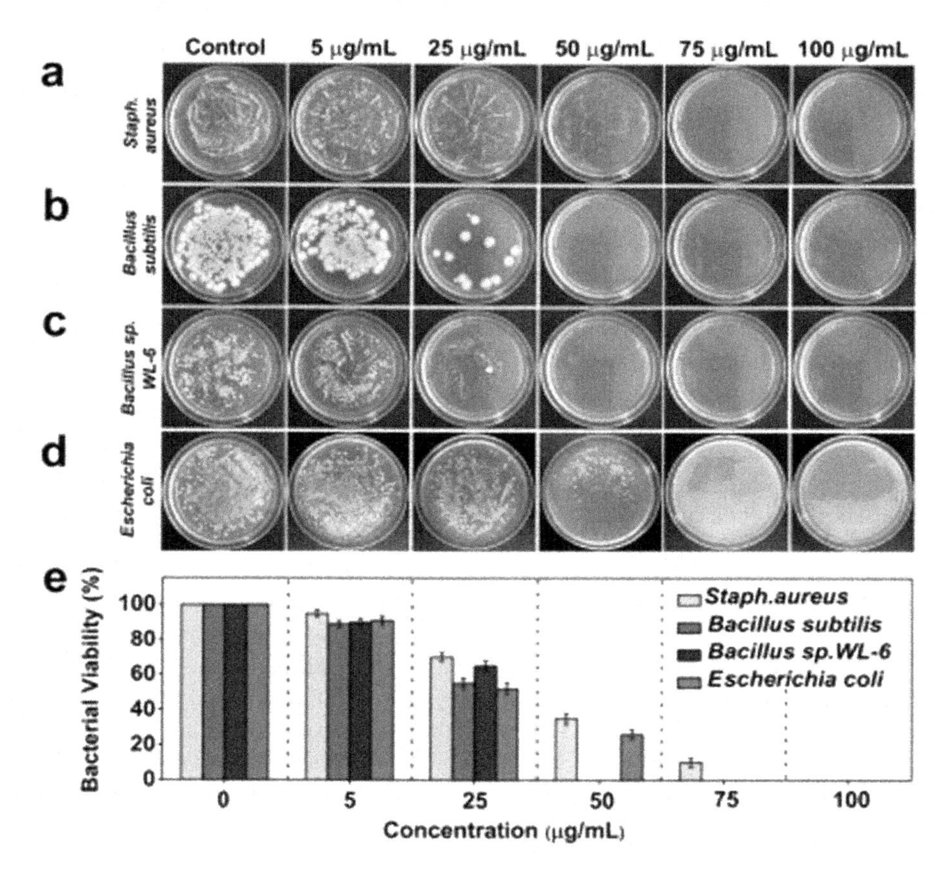

FIGURE 3.2 (a–d) Images of *Staphylococcus aureus*, *Bacillus subtilis*, *Bacillus* sp. WL-6, and *Escherichia coli* after treatment with different concentrations of CDs (0, 5, 25, 50, 75, or 100 μg/mL) for 24 h. (e) Bacterial viability was evaluated with different concentrations of CDs by UV-Vis spectroscopy. Reprinted with permission from (Li et al. 2018) © 2018 American Chemical Society.

method employing ascorbic acid (vitamin C) as a precursor elicit broad-spectrum antibacterial activity against *Staphylococcus aureus*, *Bacillus subtilis*, *Bacillus* sp. WL-6, *E. coli*, and the ampicillin-resistant *E. coli* (Figure 3.2) (Li et al. 2018). Also, CDs prepared by using ciprofloxacin hydrochloride as a precursor were found to have greater antibacterial properties against Gram-negative bacteria (Hou et al. 2017). However, CDs prepared from metronidazole as a precursor showed antibacterial activity against obligate anaerobes (Liu et al. 2017). CDs conjugated with gold nanoparticles can act as an antibacterial agent by generating ROS and thereby causing bacterial cell damage (Tang et al. 2013).

Generally, the antibacterial activity of CDs depends on various factors, like photoexcited state properties, light activation characteristics, and surface functionalities (Dong et al. 2020). Optical absorptions of CDs are responsible for eliciting the antimicrobial properties in the presence of visible light. EDA CDs under laboratory light conditions exhibited a double reduction in cell viability compared with dark conditions (Al Awak et al. 2017). Hydrophobic CDs embedded in polymer composite surfaces generated ROS in response to blue light and exerted antibacterial activity against *S. aureus*, *E. coli*, and *Klebsiella pneumoniae*. One of the disadvantages associated with using blue light is that the light itself can exert antibacterial activity towards the superficial layers of infected tissue due to poor penetrability (Figure 3.3) (Kováčová et al. 2018). In addition, CDs are known for their multiphoton absorption in the near-infrared region which can serve as an alternative to longer wavelength light activation in eliciting antibacterial properties (Cao et al. 2007; Kuo et al. 2016)

Surface functionalization of CDs is another crucial factor for antibacterial properties. Surface functional groups and charges can interact with the bacterial surface to elicit antibacterial properties (Abu Rabe et al. 2019). CDs synthesized from three different functional moieties, namely spermine for positively charged, candle soot-derived for negatively charged, and glucose for neutral, were evaluated for surface inhibition of *E. coli*. It was observed that positively charged CDs elicited greater bacterial apoptosis than did negatively charged CDs. However, no effect was observed for neutral uncharged CDs due to their inability to generate ROS (Bing et al. 2016). Another factor determining the antibacterial activity is the thickness of the surface coating of the CDs. CDs functionalized with polyethylamine of two different molecular masses, 600 and 1200 Da, were evaluated for antibacterial effect. Polyethylamine of 600 molecular mass achieved a greater reduction in viable cell number for *B. subtilis* than did polyethylamine of 1200 molecular mass. The thinner surface coating layers allow ROS to interact with bacterial cells more effectively (Abu Rabe et al. 2019).

Modification of the surface of CDs with other substances can enhance the interaction with bacterial cells. CDs decorated with TiO_2 showed antibacterial activity against *E. coli* and *S. aureus* (Yan et al. 2019). Studies were carried out on the incorporation of CDs into hydrogel to generate antibacterial surfaces (Li et al. 2018). Folic acid-conjugated polydopamine hydrogel with carbon–ZnO hybrid nanoparticles was incorporated into a wound dressing to accelerate the healing process because of its antibacterial activity (Xiang et al. 2019). Furthermore, CDs conjugated with conventional antibiotics showed increased antimicrobial properties (Dong et al. 2020).

3.5.2 Antifungal Agents

Various studies have reported the antifungal properties of CDs against the parasitic fungi *Rhizoctonia solani*, *Pyricularia grisea* (*Magnaporthe grisea*), and *Candida albicans* (Priyadarshini et al. 2018; Li et al. 2018; Jhonsi et al. 2018). On yeast cells, CDs conjugated with ciprofloxacin show a bright fluorescence owing to the presence of CDs inside the cells (Thakur et al. 2014). In another study, CDs were used to inhibit the growth of yeast cells, which was confirmed by altered cell morphology (Bagheri et al. 2018).

3.5.3 Antiviral Agents

Several studies on the antiviral activity of CDs have been reported. A recent study has shown that CD-treatment of porcine kidney cells and monkey kidney cells could

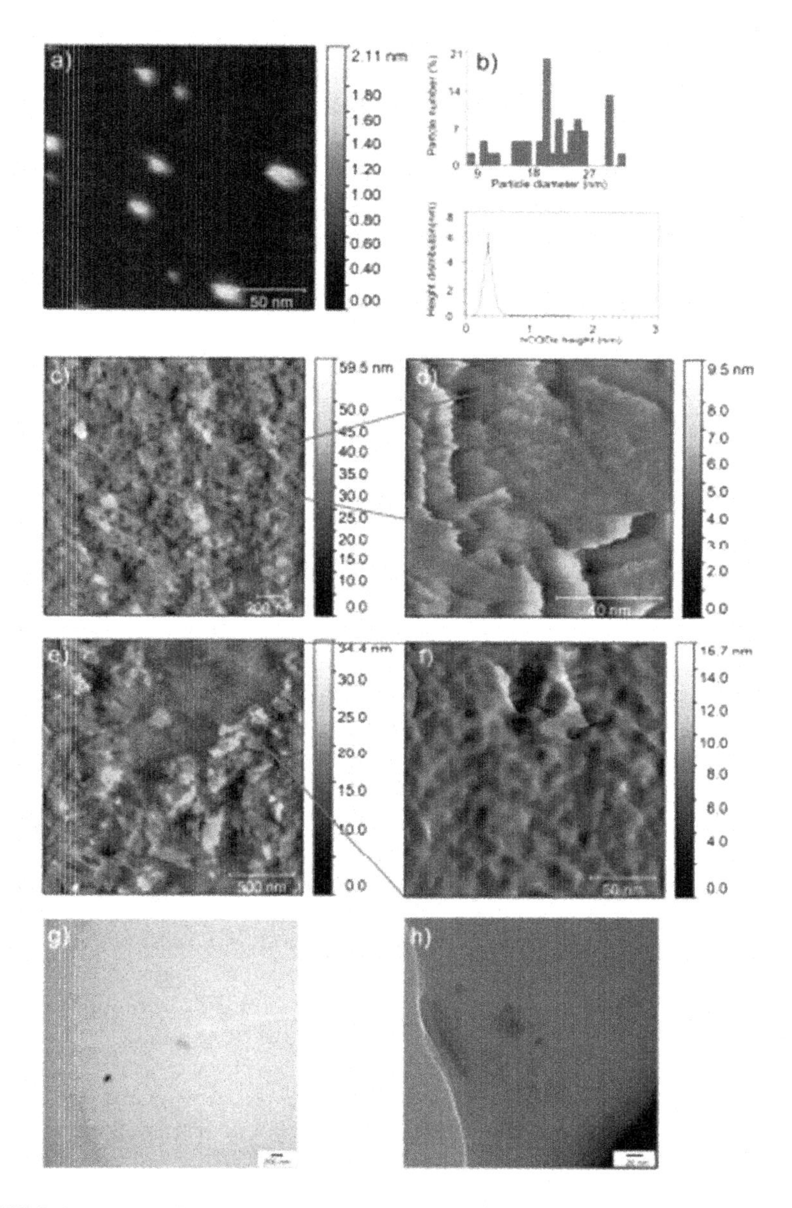

FIGURE 3.3 Atomic force microscope (AFM) images of the a) hydrophobic quantum dots (hCQDs) and b) particle size and height distributions of hCQDs. AFM images of surface morphology of c) pure Polyurethane (PU) d) inset of PU surface; e) hCQDs/PU nanocomposites; f) inset of hCQDs/PU nanocomposite surface. Transmission electron micrographs of the hCQDs/PU nanocomposites under g) low and h) high magnification. Reprinted with permission from Kováčová et al. (2018) © 2018 American Chemical Society.

inhibit the replication of the pseudorabies virus and the porcine reproductive and respiratory syndrome virus. The inhibition of virus replication involves the activation of interferons, which are innate antiviral immunity molecules (Du et al. 2016). EDA CDs show an inhibitory effect against a virus-like particle that binds to histo-blood group antigens on human cells (Dong et al. 2017). CDs prepared from benzoxamine monomer could inhibit infection *in vitro* by flaviviruses and non-enveloped viruses by direct attachment to the surface of the virion and finally inhibiting replication (Huang et al. 2019).

3.5.4 Theranostic

CDs serve as a theranostic platform by combining the diagnostic characteristics with therapeutic effects. CDs for an antimicrobial therapeutic application show concurrent effects of bioimaging, drug delivery, and photodynamic therapy (Dong et al. 2020). ZnO nanorod surface passivated with CDs elicits antibacterial and bioimaging characteristics (Mitra et al. 2012). CDs can be simultaneously used as tracers employing photoluminescent behaviour and as a drug delivery vehicle to deliver therapeutic agents, such as ciprofloxacin, as a potential theranostic (Thakur et al. 2014). CDs with antibacterial and gene delivery capability show greater efficiency than naked DNA delivery (Dou et al. 2015). Conjugation of a hybrid component of fluorescent CDs, magnetic nanoparticles, and pardaxin antimicrobial peptides can serve to achieve diagnosis, early separation, and disinfectant for a pathogen (Pramanik et al. 2017). In addition, CDs nanostructures can serve as contrast probes for two-photon imaging in a three-dimensional biological environment (Kuo et al. 2018).

The tumour imaging characteristics of CDs can be coupled with surface functionalization for therapeutic applications. Folic acid-conjugated magnetic nanoparticles decorated with CDs can be used for tumour imaging and Cu^{2+} sensing, which helps detect various pathological conditions (Kumar et al. 2016). CDs prepared from folic acid can act as a tumour imaging agent and target cancer cells (Bhunia et al. 2016). CDs prepared from vancomycin by the hydrothermal thermal method helps in pH-dependent delivery of flutamide due to a surface functional group. In addition, they can act as a cancer imaging agent (D'souza et al. 2016b). Nitrogen-doped CDs prepared from shrimps help in the triggered release of the drug boldine by the tumour pH, in addition to cellular imaging (D'souza et al. 2016a). CDs prepared by calcination of used green tea shows enhanced anticancer activity against breast cancer cell lines. Moreover, the pH-dependent variation in photoluminescence intensity can play a crucial role as a pH sensor (Hsu et al. 2013).

3.5.5 Drug Delivery

The biocompatible characteristics of CDs make them a safe drug delivery vehicle with specific modifications (Jaleel and Pramod 2018). Various studies have reported the delivery of anticancer drugs using CDs. CDs fabricated from urea and citric acid contain carboxyl groups on the surface, conjugating with the chemotherapy drug doxorubicin (DOX) by a pH-sensitive linkage. The CD–drug bond is broken in the

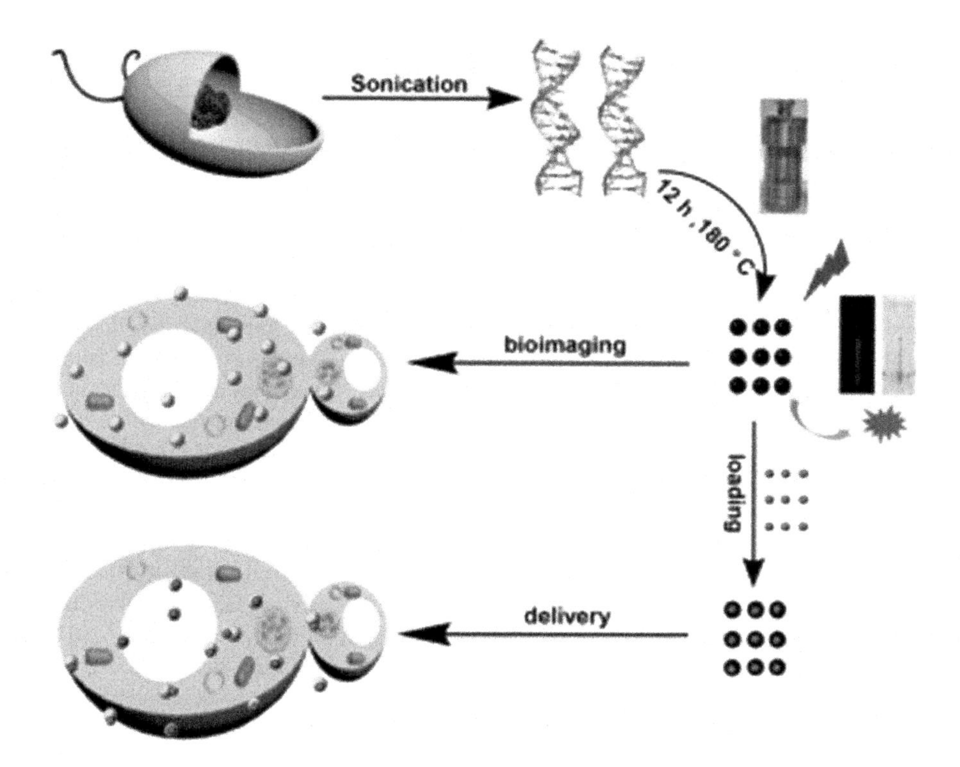

FIGURE 3.4 DNA-CDs synthesis and application in drug delivery and bioimaging. Reprinted with permission from Ding et al. (2015) © 2018 American Chemical Society.

tumour environment and releases DOX, resulting in tumour cell death (Zeng et al. 2016). The drug is released in a pH-regulated pattern by DOX loaded into hollow CDs.The empty hollow CDs without DOX were shown to be biocompatible during the CCK-8 assay. At the same time, the cytotoxic properties of DOX were enhanced with a rise in the concentration of CDs (Wang et al. 2013). Furthermore, CDs prepared using DNA from *E. coli* can be used as a drug delivery vehicle for DOX (Figure 3.4) (Ding et al. 2015). In addition, surface functionalization of CDs with folic acid promotes the receptor-mediated uptake of CDs by tumour cells, resulting in drug release in the tumour environment (Wang et al. 2013).

CDs with attached mesoporous silica nanoparticle-loaded DOX elicit low cytotoxicity against cancer cells (Zhou et al. 2013). Oligoethylenimine–β-cyclodextrin CDs complexed with hyaluronic acid-loaded DOX is found to be effective against H1299 lung cancer cells (Yang et al. 2015). Negatively charged CDs prepared from citric acid and *o*-phenylenediamine can combine with positively charged DOX with high cytocompatibility (Beibei Wang et al. 2016). CDs transferrin-DOX covalent conjugate helps to treat paediatric brain tumours (Li et al. 2016).

Conjugation of cisplatin-loaded CDs with a pH-dependent charge reversal polymer facilitates drug release at tumour cells. In a tumour environment, the anionic polymer turns cationic, resulting in the release of the positive moiety. Finally, the cisplatin pro-drug becomes activated in the reductive state and releases the drug (Figure 3.5) (Feng et al. 2016). Studies reported that lisinopril-loaded CDs cause pH-controlled drug release in lysosomes by the endocytosis mechanism (Mehta et al. 2017). Oxaliplatin-loaded CDs were prepared by surface functionalization with the amino group by a chemical coupling reaction and elicit anticancer effects (Zheng et al. 2014). CDs prepared by conjugation of $mSiO_2$ –polyethylene glycol (PEG) on the surface aids in the delivery of DOX to Hela cells (Lai et al. 2012).

The solubility of drugs is a major factor in the development of controlled-release formulations. A copolymer of chitosan and polymethyl methacrylate can be used for the preparation of a biocompatible carrier for poorly soluble drugs, which allows efficient loading of the drug into the hydrophobic core. Furthermore, conjugation of CDs to the hydrophilic surface of the drug carrier can enhance release of the drug (Chowdhuri et al. 2015). Real-time monitoring of the drug-release profile, based on the Forster resonance energy transfer (FRET) signal, can be achieved by fabricating a surface-engineered FRET-based CDs system. Such a system for delivery of DOX could be achieved by surface functionalization of PEGylated CDs with folic acid, where the signal was observed at 498 nm after drug release (Tang et al. 2013).

Moreover, high loading of DOX could be achieved by surface functionalization of CDs prepared from sorbitol by bovine serum albumin (Mewada et al. 2014). Recently, a drug delivery system was prepared between ion liquid-mediated organophilic CDs and curcumin (Cur), using hydrophobic interactions to be efficient drug carriers, with a rapid cell-penetrating potential and induction of apoptosis against HeLa cells (Shu et al. 2017). The addition of functional groups onto the surface of CDs helps in the targeted delivery of drugs (Mishra et al. 2018). A nucleus-targeted drug delivery system, using zwitterionic CDs with grafted DOX, helps in anticancer therapy, with high biocompatibility (Jung et al. 2015). Nuclear-localized single peptide CDs for DOX show greater efficiency in killing cancer cells, compared with free DOX (Yang et al. 2016). Recently, amine-functionalized DNA CDs hydrogel helps in the sustained and targeted release of DOX, where the CDs play dual functions as crosslinkers and drug-encapsulating agents (Singh et al. 2017).

3.5.6 Gene Therapy

The incorporation of CDs into gene delivery vectors will give a basic visualization for image-guided gene therapy. Passivation of CDs derived from citric acid with polyethylamine helps to condense surviving small interfering RNA (siRNA), thereby forming a gene transfection vector in addition to a bioimaging agent (Wang et al. 2014). It was observed that CDs containing vectors have high transfection efficiency (Liu et al. 2012). Moreover, CDs derived from branched polyethyleneimine (PEI) can act as a gene delivery vector. Some unchanged branches on the surface of oxidized branched PEI are responsible for gene delivery (Hu et al. 2014). Recently, hyaluronic acid–PEI-functionalized CDs show high biocompatibility with gene condensation ability, making them crucial for gene delivery (Zhang et al. 2017). Folate-conjugated reducible PEI-passivated CDs act as a siRNA gene carrier in lung

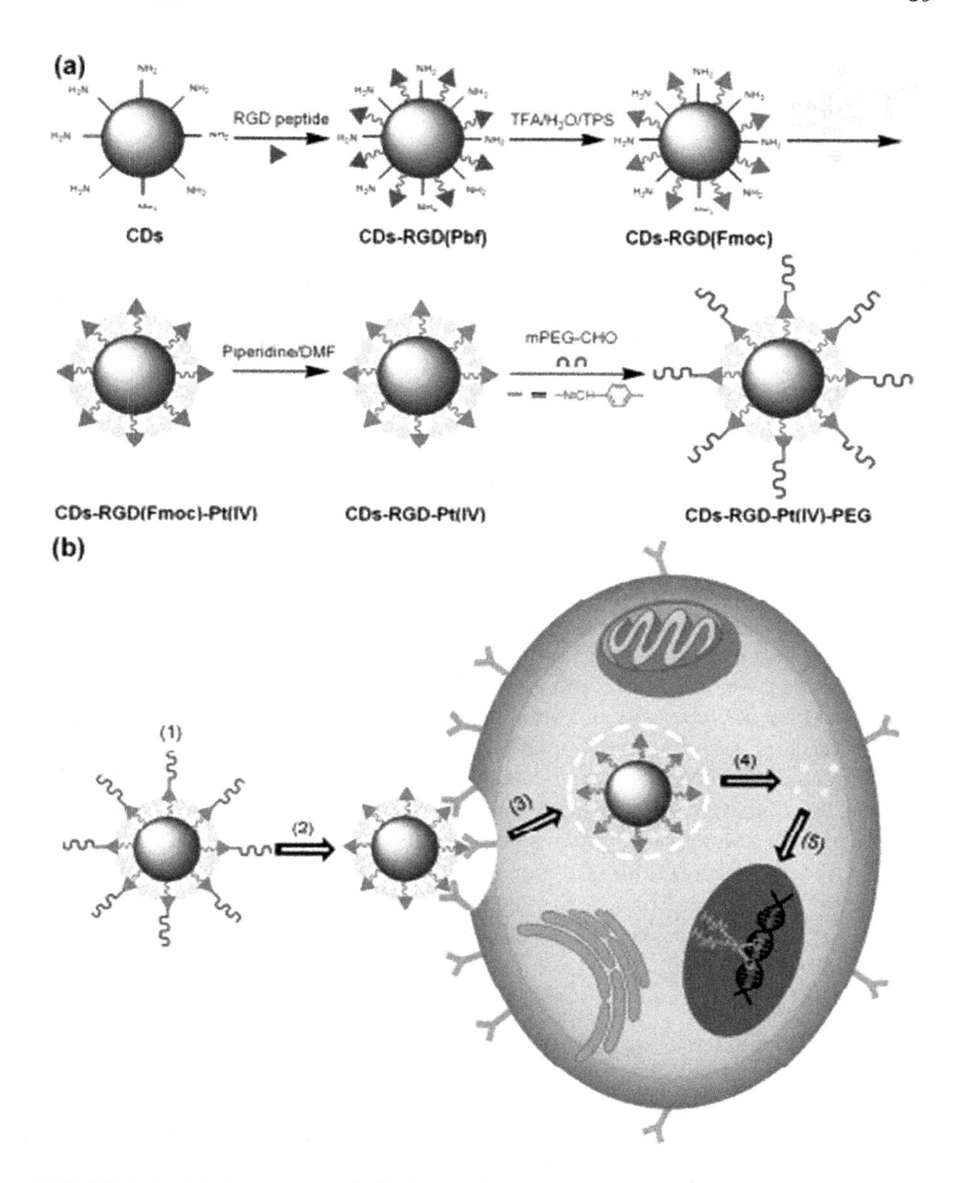

FIGURE 3.5 (a) Preparation of CD-based drug nanocarrier CDs-RGD-Pt(IV)-PEG with tumour-triggered targeting property. (b) Schematic illustration of the drug delivery process using CDs-RGD-Pt(IV)-PEG: (1) tracking drug nanocarriers by multicolour fluorescence of CDs with PEGylation under normal physiological condition; (2) tumour-triggered targeting ligands exposed at tumour extracellular pH 6.8; (3) effective uptake by cancer cells through ligand–receptor interaction; (4) cisplatin release from cisplatin(IV) pro-drug under reductive cytosol; and (5) cisplatin binding with DNA to exhibit the cytotoxicity. Reprinted with permission from Feng et al. (2016) © 2016 American Chemical Society.

cancer gene therapy (Wu et al. 2016). Furthermore, fluorine-doped carbon atoms, prepared by a ring-opening reaction, show high gene transfection efficiency and cellular uptake, making them excellent gene delivery vectors (Luo et al. 2018).

Interestingly, folic acid-derived CDs were suggested to be useful for tumour-targeted delivery of therapeutic agents and diagnosis (Bhunia et al. 2016). An extension of this application was demonstrated by attaching folic acid-derived CDs onto diamine PEG-passivated graphene-chitosan composites for targeted delivery of the Tumor Necrosis Factor-Alpha (TNF-α) gene to tumour cells (Figure 3.6) (Jaleel et al. 2019).

3.5.7 Detection of Drugs

CDs can play a role as detection probes for drugs and other molecules. The fluorescence intensity decreases when tetracycline binds to the surface groups of CDs. When CDs prepared from L-cysteine and diphosphorus pentoxide by the non-thermal method are employed, tetracycline at a concentration below 7.5 nM can be detected in a urine sample and pharmaceutical preparations (Yang et al. 2014). In addition, the presence of catechol groups on CDs prepared from tannic acid can be used to detect oxytetracycline (An et al. 2015). The detection of illicit drugs, using spectroscopic and chromatographic techniques, is highly complex and expensive. Therefore, the photoluminescence property of CDs prepared from a mixture of citric acid and urea can be employed for the detection of illicit drug precursors such as methamphetamine precursors, amphetamine sulphate, etc. (Kim et al. 2015). Fluorescent CDs, prepared by carbonization of tea powder, helps in the detection of dopamine and ascorbic acid at concentrations of less than 33 and 98 μM, respectively, where fluorescence decreases in response to an increase in the concentration of the molecule to be detected (Baruah et al. 2014). CDs synthesized from papain can act as a sensor for determining hydrogen peroxide, doxycycline, and iodine when used in combination with papain template gold nanoclusters (Yang et al. 2017).

3.5.8 Food Quality and Safety

CDs play a significant role in the area of food quality and safety, such as in the detection of nutrients, of restricted or banned items, and determining the presence of pathogenic bacteria and toxins secreted by them, as well as quantification of the amount of pesticides with high sensitivity and selectivity (Qu et al. 2018). One of the important criteria for various food applications is that they must be non-cytotoxic.

The MnO_2-CDs fluorescence sensing system helps determine ascorbic acid in fruits, vegetables, and fresh juices with a limit of detection of 42 nm. The fluorescence intensity of CDs is quenched when MnO2 nanosheets are introduced. When ascorbic acid is introduced, the quenching is removed and fluorescence is retained (Liu et al. 2016). Fluorescent CDs prepared by thermal carbonization is used to determine tannic acid with a limit of detection of 0.018 mg L^{-1} (Ahmed et al. 2015). CDs prepared by the hydrothermal method, using aloe vera as a carbon source, can

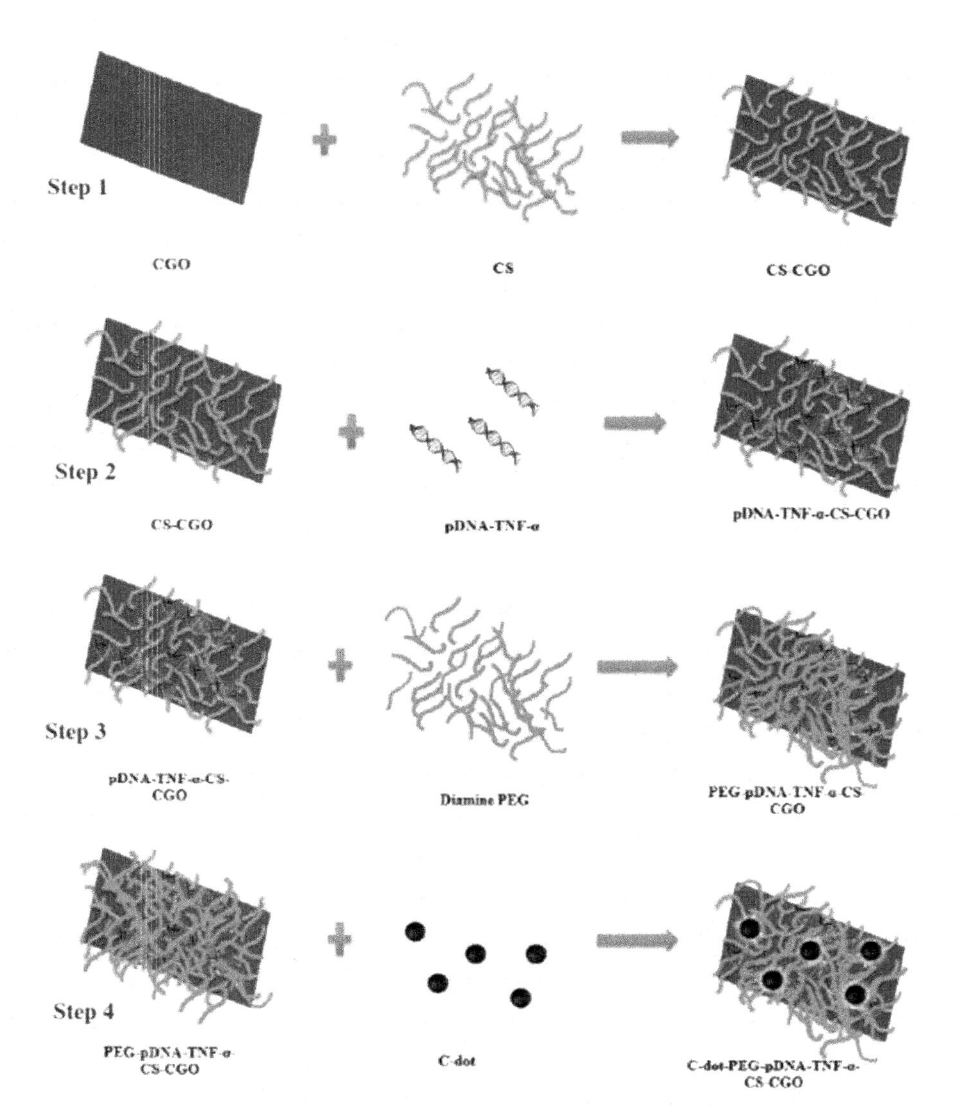

FIGURE 3.6 Steps involved in the preparation of folic acid-derived carbon dots (CDs) attached to diamine poly(ethylene glycol) (PEG)-passivated graphene-chitosan composite for targeted delivery of Tumour Necrosis Factor (TNF)-α gene to tumour cells [CGO = carboxylated graphene oxide, CS = Chitosan]. Reprinted with permission from Jaleel et al. (2019) © 2019 Elsevier.

be used to determine the concentration of the synthetic food colourant tartrazine in food samples (Xu et al. 2015).

The problem associated with pesticide contamination is another severe concern regarding food safety. The fluorescence-based immune reaction method was employed to determine glyphosate concentrations in Pearl River Water and tea samples, with a limit of detection of 8 ng/mL. Antibody-labelled CDs were allowed to mix with antigen-coated magnetic beads. Glyphosate concentration can be determined by removing antigen-coated magnetic beads by magnetic separation (Wang et al. 2016).

In addition, CDs play a crucial role in the detection of microorganisms and toxins with high specificity. A combination of a CDs and an aptamer can act as a fluorescent probe for the quantitative determination of *Salmonella typhimurium*. CDs prepared by the hydrothermal method using citric acid as a precursor is combined with amino-modified aptamers to ensure high specificity. This rapid and sensitive method helped determine *S. typhimurium* egg samples and tap water with a detection limit of 50 colony-forming units (CFU)/mL (Wang et al. 2015). Novel nitrogen-doped CDs prepared by the hydrothermal method, using pancreatin as a precursor, can be assembled on gold nanoparticles by electrostatic attachment to detect aflatoxin B1 with a detection limit of 5 pg/mL(Wang et al. 2016).

3.5.9 Bioimaging

The photoluminescence emission property of CDs in the near-infrared region plays the main role during *in-vivo* bioimaging applications (Namdari et al. 2017). Figure 3.7 represents the general mechanism of drug delivery using CDs and the bioimaging process. The unique characteristics, like photostability, photobleaching resistance,

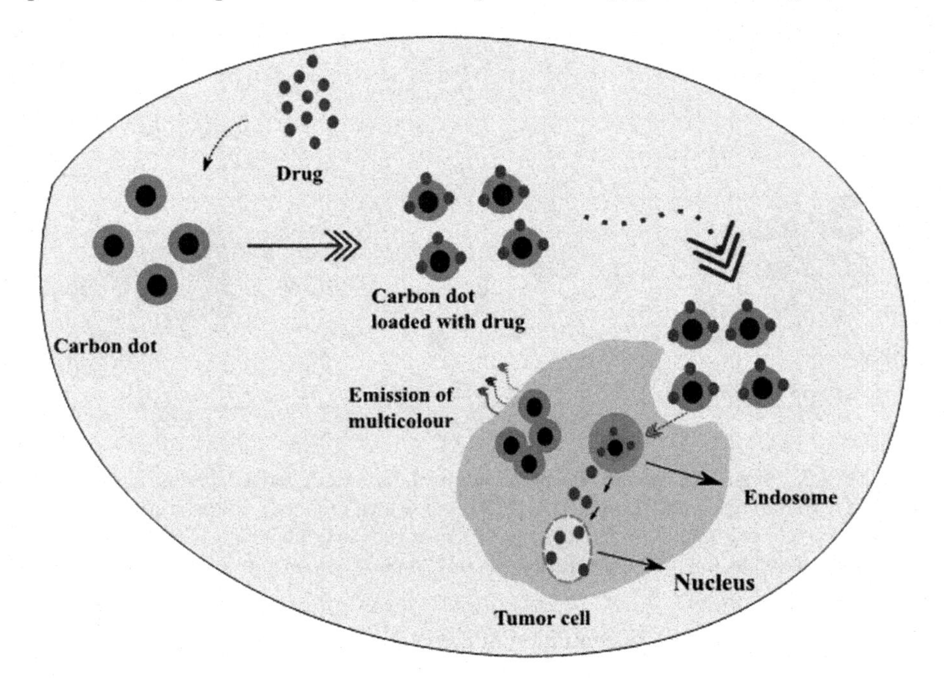

FIGURE 3.7 The mechanism of drug delivery using CDs. The drug-loaded CDs enter into the cell *via* endocytosis and deliver the drug into the nucleus. The multicoloUr emission of CDs inside the cell helps to achieve cellular imaging.

low cytotoxicity, and two-photon emission properties, make them promising bioimaging probes. The fluorescence emission property of CDs is considered in the development of nanoprobes for bioimaging applications. In addition, the particle size of CDs nanoprobes can be controlled by tuning the carbonization conditions. Moreover, surface passivation can enhance photostability with different polymers (Jaleel and Pramod 2018). The fluorescence ability of CDs helps to differentiate between normal and cancerous cells as tumour cells have greater permeability due to cell membrane alterations (Dekaliuk et al. 2015).

In-vivo imaging of plant cells is more complex than that of animal cells due to the complicated structure of plant cells. Studies with N- and S-doped CDs show high potential for *in-vivo* imaging of plant cells. Chemical modification of CDs prepared from waste carbon sources with ethylenediamine can act as live-cell imaging probes for breast cancer cells (Angelis et al. 2015). N- and P-CDs prepared by microwave-assisted thermolysis can act as a potential nanoprobes for cellular imaging with high stability (Li et al. 2016).

Nitrogen-doped CDs, which are prepared from streptomycin by the hydrothermal method and show good biocompatibility, exhibit bright blue fluorescence in cell imaging (Wang et al. 2014). Passivated CDs were found to have great applications for *in-vivo* photoacoustic imaging of sentinel lymph nodes. The sudden response was observed in the infrared region due to their small size and rapid lymphatic transport, compared with dyes (Wu et al. 2013). CDs prepared from pomegranate (*Punica granatum*) fruits can serve as an imaging agent for bacterial, fungal, plant, and animal cells with no cytotoxicity (Kasibabu et al. 2015). Fluorescence imaging studies conducted with zebrafish embryos demonstrated the potential of the highly biocompatible nature of CDs (Kang et al. 2015).

Nanocomposites prepared by incorporating CDs into silica nanorattle can act as *in-vitro* cell labelling and *in-vivo* imaging agents. It was observed that the presence of silica nanorattle enhances the fluorescence intensity to higher than that achieved by CDs alone (Fu et al. 2015). *Ex-vivo* imaging of guppy fish can be carried out with plant soot-derived and Nescafe-derived CDs.

3.5.10 Biosensing

Because of their fluorescent properties, CDs are most widely used for biosensing purposes. They are used for the detection of metal ions, anions, and several biomolecules (Zhang and Yu 2016). Several biological and chemical sensors have been developed using them (Tuerhong et al. 2017).

In a study, the CDs were prepared by microwave pyrolysis, and the synthesized CDs were added to 1% chitosan solution to obtain modified chitosan CDs. Using *p*-phenylenediamine as a carbon source, N-CDs with bright orange fluorescence emission at 592 nm were synthesized using the microwave method.They show excellent solubility in water, high biocompatibility, optical stability, and pH response behaviour.They show a linear response to the pH range of 4.45 to 7.0. Later, they were used as a platform for the detection of vitamin D2 (Sarkar et al. 2018). Hence, they were used as a fluorescent probe for cellular imaging and for detecting changes in the pH of living cells (Ding et al. 2018). Cobalt-doped CDs were used for the detection of L-cysteine in human serum. Specific recognition of cobalt-doped CDs enables the detection of L-cysteine with greater sensitivity and selectivity, without the interference from glutathione or homocysteine (Liu et al. 2019).

The *Saccharomyces*-derived N-CDs have been used as probes for detecting vitamin B_{12} and sensing changes in pH (Yu et al. 2019). The manganese oxide-doped CDs exhibits *in-vitro* thermosensitive properties within the temperature range 10–60°C. Hence, they can be used as a thermometer in living cells. Interestingly, studies revealed that this compound could be used as an adjuvant for the treatment of liver cancer (Li et al. 2020).

In a study, CDs were synthesized from gadong (*Dioscorea hispida*) tubers by acid hydrolysis, acid dehydration, and oxidation steps, and was later used to detect arginine (Sonthanasamy et al. 2020). Cancer marker detection and analysis of tumour cells is a difficult task to achieve. Several nanomaterials were used for this purpose. Recent investigations have revealed the potential of CDs tp sense cancer markers and analyse tumour cells. The proper functionalization of CDs with a suitable molecule will help to target a particular part of the cell. Further investigations are to be conducted to explore the potential of CDs as an efficient platform for biosensing (Pirsaheb et al. 2019).

3.6 CONCLUSIONS

CDs are an emerging class of carbon nanoparticles, displaying a wide range of fascinating properties. This chapter focused on the structure, properties, various top-down and bottom-up synthesis methods, and health care applications of CDs. CDs have a wide range of applications in bioimaging, biosensing, drug delivery, catalysis, and other health care fields. The fluorescent nature and biocompatibility of CDs enable their use in both *in-vitro* and *in-vivo* bioimaging and for several diagnostic purposes. The doping of CDs with atoms like nitrogen and sulphur was done to enhance their applications. Wide attention is now focused on CDs because of their high luminescence and water solubility. Recent studies have achieved important steps forward in the current understanding of CDs by developing a range of advanced methods to understand the unique properties of CDs. Despite all these advances, more research is needed. In particular, the photophysical behaviour and the structure-controlled synthesis are still beyond our reach. Further studies are needed in this field to overcome all those challenges associated with the properties and applications of CDs.

3.7 REFERENCES

Abu Rabe, Dina I., Mohamad M. Al Awak, Fan Yang, Peter A. Okonjo, Xiuli Dong, Lindsay R. Teisl, Ping Wang, et al. 2019. "The Dominant Role of Surface Functionalization in Carbon Dots' Photo-Activated Antibacterial Activity." *International Journal of Nanomedicine* 14 (April). Dove: 2655–65. doi:10.2147/IJN.S200493.

Ahirwar, Satyaprakash, Sudhanshu Mallick, and Dhirendra Bahadur. 2017. "Electrochemical Method To Prepare Graphene Quantum Dots and Graphene Oxide Quantum Dots." *ACS Omega* 2 (11). American Chemical Society: 8343–53. doi:10.1021/acsomega.7b01539.

Ahmed, Gaber Hashem Gaber, Rosana Badía Laíño, Josefa Angela García Calzón, and Marta Elena Díaz García. 2015. "Fluorescent Carbon Nanodots for Sensitive and Selective Detection of Tannic Acid in Wines." *Talanta* 132: 252–57. doi:https://doi.org/10.1016/j.talanta.2014.09.028.

An, Xuting, Shujuan Zhuo, Ping Zhang, and Changqing Zhu. 2015. "Carbon Dots Based Turn-on Fluorescent Probes for Oxytetracycline Hydrochloride Sensing." *RSC Advances* 5 (26). The Royal Society of Chemistry: 19853–58. doi:10.1039/C4RA16456C.

Angelis, Cintya D., E. S. Barbosa, Josø R. Corrþa, Gisele A. Medeiros, Gabrielle Barreto, Kelly G. Magalhes, Aline L. De Oliveira, John Spencer, Marcelo O. Rodrigues, and Brenno A. D. Neto. 2015. "Carbon Dots (C-Dots) from Cow Manure with Impressive Subcellular Selectivity Tuned by Simple Chemical Modification." *Chemistry: A European Journal* 21: 5055–60. doi:10.1002/chem.201406330.

Anwar, Sadat, Haizhen Ding, Mingsheng Xu, Xiaolong Hu, Zhenzhen Li, Jingmin Wang, Li Liu, et al. 2019. "Recent Advances in Synthesis, Optical Properties, and Biomedical Applications of Carbon Dots." *ACS Applied Bio Materials* 2 (6). American Chemical Society: 2317–38. doi:10.1021/acsabm.9b00112.

Arora, Neha, and N. N. Sharma. 2014. "Arc Discharge Synthesis of Carbon Nanotubes: Comprehensive Review." *Diamond and Related Materials* 50: 135–50. doi:https://doi.org/10.1016/j.diamond.2014.10.001.

Awak, Mohamad M. Al, Ping Wang, Shengyuan Wang, Yongan Tang, Ya-Ping Sun, and Liju Yang. 2017. "Correlation of Carbon Dots' Light-Activated Anti https://www.google.com/search?q=gadong+tuber&oq=gadong+tuber&aqs=chrome..69i57.3163j0j4&sourceid=chrome&ie=UTF-8#:~:text=with%20Ubi%20Gadong%20(-,Dioscorea%20hispida,-)%20Extract%20...%20is%20the ouricrobial Activities and Fluorescence Quan https://www.google.com/search?q=gadong+tuber&oq=gadong+tuber&aqs=chrome..69i57.3163j0j4&sourceid=chrome&ie=UTF-8#:~:text=with%20Ubi%20Gadong%20(-,Dioscorea%20hispida,-)%20Extract%20...%20is%20the m Yield." *RSC Advances* 7 (48). The Royal Society of Chemistry: 30177–84. doi:10.1039/C7RA05397E.

Bagheri, Zeinab, Hamide Ehtesabi, Zahra Hallaji, Neda Aminoroaya, Hossein Tavana, Ebrahim Behroodi, Mahban Rahimifard, Mohammad Abdollahi, and Hamid Latifi. 2018. "On-Chip Analysis of Carbon Dots Effect on Yeast Replicative Lifespan." *Analytica Chimica Acta* 1033: 119–27. doi:https://doi.org/10.1016/j.aca.2018.05.005.

Baruah, Upama, Neelam Gogoi, Achyut Konwar, Manash Jyoti Deka, Devasish Chowdhury, and Gitanjali Majumdar. 2014. "Carbon Dot Based Sensing of Dopamine and Ascorbic Acid." Edited by Shu Taira. *Journal of Nanoparticles* 2014. Hindawi Publishing Corporation: 178518. doi:10.1155/2014/178518.

Berenguel-Alonso, M., I. Ortiz-Gómez, B. Fernández, P. Couceiro, J. Alonso-Chamarro, L F. Capitán-Vallvey, A. Salinas-Castillo, and M. Puyol. 2019. "An LTCC Monolithic Microreactor for the Synthesis of Carbon Dots with Photoluminescence Imaging of the Reaction Progress." *Sensors and Actuators B: Chemical* 296: 126613. doi:https://doi.org/10.1016/j.snb.2019.05.090.

Bhartiya, Prabha, Anu Singh, Hridyesh Kumar, Tanvi Jain, Brijesh Singh, and Pradip Dutta. 2016. "Carbon Dots : Chemistry, Properties and Applications." *Indian Chemical Society* 93 (July): 759–66.

Bhunia, Susanta Kumar, Amit Ranjan Maity, Sukhendu Nandi, David Stepensky, and Raz Jelinek. 2016. "Imaging Cancer Cells Expressing the Folate Receptor with Carbon Dots Produced from Folic Acid." *ChemBioChem* 17 (7). Wiley: 614–19. doi:10.1002/cbic.201500694.

Bing, Wei, Hanjun Sun, Zhengqing Yan, Jinsong Ren, and Xiaogang Qu. 2016. "Programmed Bacteria Death Induced by Carbon Dots with Different Surface Charge." *Small* 12 (34). Wiley: 4713–18. doi:10.1002/smll.201600294.

Cao, Li, Xin Wang, Mohammed J. Meziani, Fushen Lu, Haifang Wang, Pengju G. Luo, Yi Lin, et al. 2007. "Carbon Dots for Multiphoton Bioimaging." *Journal of the American Chemical Society* 129 (37). American Chemical Society: 11318–19. doi:10.1021/ja0735271.

D'souza, Stephanie L., Balaji Deshmukh, Jigna R. Bhamore, Karuna A. Rawat, Nibedita Lenka, and Suresh Kumar Kailasa. 2016a. "Synthesis of Fluorescent Nitrogen-Doped Carbon Dots from Dried Shrimps for Cell Imaging and Boldine Drug Delivery System." *RSC Advances* 6 (15). The Royal Society of Chemistry: 12169–79. doi:10.1039/C5RA24621K.

D'souza, Stephanie L., Balaji Deshmukh, Karuna A. Rawat, Jigna R. Bhamore, Nibedita Lenka, and Suresh Kumar Kailasa. 2016b. "Fluorescent Carbon Dots Derived from Vancomycin for Flutamide Drug Delivery and Cell Imaging." *New Journal of Chemistry* 40 (8). The Royal Society of Chemistry: 7075–83. doi:10.1039/C6NJ00358C.

Dager, Akansha, Takashi Uchida, Toru Maekawa, and Masaru Tachibana. 2019. "Synthesis and Characterization of Mono-Disperse Carbon Quantum Dots from Fennel Seeds: Photoluminescence Analysis Using Machine Learning." *Scientific Reports* 9 (1). Nature Publishing Group UK: 14004. doi:10.1038/s41598-019-50397-5.

Dekaliuk, Mariia, Kyrylo Pyrshev, and Alexander Demchenko. 2015. "Visualization and Detection of Live and Apoptotic Cells with Fluorescent Carbon Nanoparticles." *Journal of Nanobiotechnology* 13 (1): 86. doi:10.1186/s12951-015-0148-7.

Ding, Han, Feiyue Du, Pengchang Liu, Zhijun Chen, and Jiacong Shen. 2015. "DNA–Carbon Dots Function as Fluorescent Vehicles for Drug Delivery." *ACS Applied Materials & Interfaces* 7 (12). American Chemical Society: 6889–97. doi:10.1021/acsami.5b00628.

Ding, Yuan-Yuan, Xiao-Juan Gong, Yang Liu, Wen-Jing Lu, Yi-Fang Gao, Ming Xian, Shao-Min Shuang, and Chuan Dong. 2018. "Facile Preparation of Bright Orange Fluorescent Carbon Dots and the Constructed Biosensing Platform for the Detection of PH in Living Cells." *Talanta* 189: 8–15. doi:https://doi.org/10.1016/j.talanta.2018.06.060.

Dong, Xiuli, Weixiong Liang, Mohammed J. Meziani, Ya-ping Sun, and Liju Yang. 2020. "Carbon Dots as Potent Antimicrobial Agents." *Theranostics* 10 (2): 671–86. doi:10.7150/thno.39863.

Dong, Xiuli, Marsha M. Moyer, Fan Yang, Ya-Ping Sun, and Liju Yang. 2017. "Carbon Dots' Antiviral Functions Against Noroviruses." *Scientific Reports* 7 (1): 519. doi:10.1038/s41598-017-00675-x.

Dou, Qingqing, Xiaotian Fang, Shan Jiang, Pei Lin Chee, Tung-Chun Lee, and Xian Jun Loh. 2015. "Multi-Functional Fluorescent Carbon Dots with Antibacterial and Gene Delivery Properties." *RSC Advances* 5 (58). The Royal Society of Chemistry: 46817–22. doi:10.1039/C5RA07968C.

Du, Ting, Jiangong Liang, Nan Dong, Lin Liu, Liurong Fang, Shaobo Xiao, and Heyou Han. 2016. "Carbon Dots as Inhibitors of Virus by Activation of Type I Interferon Response." *Carbon* 110: 278–85. doi:https://doi.org/10.1016/j.carbon.2016.09.032.

Feng, Tao, Xiangzhao Ai, Guanghui An, Piaoping Yang, and Yanli Zhao. 2016. "Charge-Convertible Carbon Dots for Imaging-Guided Drug Delivery with Enhanced in Vivo Cancer Therapeutic Efficiency." *ACS Nano* 10 (4). American Chemical Society: 4410–20. doi:10.1021/acsnano.6b00043.

Fong, Jessica Fung Yee, Suk Fun Chin, and Sing Muk Ng. 2016. "A Unique 'Turn-on' Fluorescence Signalling Strategy for Highly Specific Detection of Ascorbic Acid Using Carbon Dots as Sensing Probe." *Biosensors and Bioelectronics* 85: 844–52. doi:https://doi.org/10.1016/j.bios.2016.05.087.

Fu, Changhui, Li Qiang, Qinghua Liang, Xue Chen, Linlin Li, Huiyu Liu, Longfei Tan, Tianlong Liu, Xiangling Ren, and Xianwei Meng. 2015. "Facile Synthesis of a Highly Luminescent Carbon Dot@silica Nanorattle for in Vivo Bioimaging." *RSC Advances* 5 (57). The Royal Society of Chemistry: 46158–62. doi:10.1039/C5RA04311E.

Gayen, Biswajit, Soubantika Palchoudhury, and Joydeep Chowdhury. 2019. “Carbon Dots: A Mystic Star in the World of Nanoscience.” Edited by Oscar Perales-Pérez. *Journal of Nanomaterials* 2019. Hindawi: 3451307. doi:10.1155/2019/3451307.

Ge, Jiechao, Minhuan Lan, Bingjiang Zhou, Weimin Liu, Liang Guo, Hui Wang, Qingyan Jia, et al. 2014. “A Graphene Quantum Dot Photodynamic Therapy Agent with High Singlet Oxygen Generation.” *Nature Communications* 5:4596. doi:10.1038/ncomms5596.

Gong, Ningqiang, Hao Wang, Shuai Li, Yunlong Deng, Xiao’ai Chen, Ling Ye, and Wei Gu. 2014. “Microwave-Assisted Polyol Synthesis of Gadolinium-Doped Green Luminescent Carbon Dots as a Bimodal Nanoprobe.” *Langmuir* 30 (36). American Chemical Society: 10933–39. doi:10.1021/la502705g.

Gul, Urooj, Shamsa Kanwal, Sobia Tabassum, Mazhar Amjad Gilani, and Abdur Rahim. 2020. “Microwave-Assisted Synthesis of Carbon Dots as Reductant and Stabilizer for Silver Nanoparticles with Enhanced-Peroxidase like Activity for Colorimetric Determination of Hydrogen Peroxide and Glucose.” *Microchimica Acta* 187 (2): 135. doi:10.1007/s00604-019-4098-x.

Han, Zhu, Li He, Shuang Pan, Hui Liu, and Xiaoli Hu. 2020. “Hydrothermal Synthesis of Carbon Dots and Their Application for Detection of Chlorogenic Acid.” *Luminescence* 35 (7). Wiley: 989–97. doi:https://doi.org/10.1002/bio.3803.

Hou, Peng, Tong Yang, Hui Liu, Yuan Fang Li, and Cheng Zhi Huang. 2017. “An Active Structure Preservation Method for Developing Functional Graphitic Carbon Dots as an Effective Antibacterial Agent and a Sensitive PH and Al(Iii) Nanosensor.” *Nanoscale* 9 (44). The Royal Society of Chemistry: 17334–41. doi:10.1039/C7NR05539K.

Hou, Yuxin, Qiujun Lu, Jianhui Deng, Haitao Li, and Youyu Zhang. 2015. “One-Pot Electrochemical Synthesis of Functionalized Fluorescent Carbon Dots and Their Selective Sensing for Mercury Ion.” *Analytica Chimica Acta* 866: 69–74. doi:https://doi.org/10.1016/j.aca.2015.01.039.

Hsu, Pin-Che, Po-Cheng Chen, Chung-Mao Ou, Hsin-Yun Chang, and Huan-Tsung Chang. 2013. “Extremely High Inhibition Activity of Photoluminescent Carbon Nanodots toward Cancer Cells.” *Journal of Materials Chemistry B* 1 (13). The Royal Society of Chemistry: 1774–81. doi:10.1039/C3TB00545C.

Hu, Liming, Yun Sun, Shengliang Li, Xiaoli Wang, Kelei Hu, Lirong Wang, Xing-jie Liang, and Yan Wu. 2014. “Multifunctional Carbon Dots with High Quantum Yield for Imaging and Gene Delivery.” *Carbon* 67: 508–13. doi:https://doi.org/10.1016/j.carbon.2013.10.023.

Hu, Shengliang, Jun Liu, Jinlong Yang, Yanzhong Wang, and Shirui Cao. 2011. “Laser Synthesis and Size Tailor of Carbon Quantum Dots.” *Journal of Nanoparticle Research* 13 (12): 7247–52. doi:10.1007/s11051-011-0638-y.

Huang, Shaomei, Jiangjiang Gu, Jing Ye, Bin Fang, Shengfeng Wan, Caoyu Wang, Usama Ashraf, et al. 2019. “Benzoxazine Monomer Derived Carbon Dots as a Broad-Spectrum Agent to Block Viral Infectivity.” *Journal of Colloid and Interface Science* 542: 198–206. doi:https://doi.org/10.1016/j.jcis.2019.02.010.

Huang, Xinglu, Fan Zhang, Lei Zhu, Ki Young Choi, Ning Guo, Jinxia Guo, Kenneth Tackett, et al. 2013. “Effect of Injection Routes on the Biodistribution, Clearance, and Tumor Uptake of Carbon Dots.” *ACS Nano* 7 (7). American Chemical Society: 5684–93. doi:10.1021/nn401911k.

Ipe, Binil Itty, Manfred Lehnig, and Christof M. Niemeyer. 2005. “On the Generation of Free Radical Species from Quantum Dots.” *Small* 1 (7). Wiley: 706–9. doi:10.1002/smll.200500105.

Jaiswal, Amit, Siddhartha Sankar Ghsoh, and Arun Chattopadhyay. 2012. “Quantum Dot Impregnated-Chitosan Film for Heavy Metal Ion Sensing and Removal.” *Langmuir* 28 (44). American Chemical Society: 15687–96. doi:10.1021/la3027573.

Jaleel, Jumana Abdul, and K. Pramod. 2018. "Artful and Multifaceted Applications of Carbon Dot in Biomedicine." *Journal of Controlled Release*. doi:10.1016/j.jconrel.2017.11.027.

Jaleel, Jumana Abdul, Shabeeba M. Ashraf, Krishnan Rathinasamy, and K. Pramod. 2019. "Carbon Dot Festooned and Surface Passivated Graphene-Reinforced Chitosan Construct for Tumor-Targeted Delivery of TNF-α Gene." *International Journal of Biological Macromolecules* 127: 628–36. doi:https://doi.org/10.1016/j.ijbiomac.2019.01.174.

Jhonsi, Mariadoss Asha, Devanesan Arul Ananth, Gayathri Nambirajan, Thilagar Sivasudha, Rekha Yamini, Soumen Bera, and Arunkumar Kathiravan. 2018. "Antimicrobial Activity, Cytotoxicity and DNA Binding Studies of Carbon Dots." *Spectrochimica Acta. Part A, Molecular and Biomolecular Spectroscopy* 196. Department of Chemistry, B. S. Abdur Rahman Crescent Institute of Science and Technology, Chennai, Tamil Nadu, India. Electronic address: asha@bsauniv.ac.in.: 295–302. doi:10.1016/j.saa.2018.02.030.

Jung, Yun Kyung, Eeseul Shin, and Byeong-Su Kim. 2015. "Cell Nucleus-Targeting Zwitterionic Carbon Dots." *Scientific Reports* 5 (1): 18807. doi:10.1038/srep18807.

Jusuf, Bella Nathanael, Nonni Soraya Sambudi, and Shafirah Samsuri. 2018. "Microwave-Assisted Synthesis of Carbon Dots from Eggshell Membrane Ashes by Using Sodium Hydroxide and Their Usage for Degradation of Methylene Blue." *Journal of Environmental Chemical Engineering* 6 (6): 7426–33. doi:https://doi.org/10.1016/j.jece.2018.10.032.

Kang, Yan-Fei, Yu-Hao Li, Yang-Wu Fang, Yang Xu, Xiao-Mi Wei, and Xue-Bo Yin. 2015. "Carbon Quantum Dots for Zebrafish Fluorescence Imaging." *Scientific Reports* 5 (1): 11835. doi:10.1038/srep11835.

Kang, Zhenhui, Yang Liu, and Shuit-Tong Lee. 2019. "Carbon Dots for Bioimaging and Biosensing Applications BT." In *Carbon-Based Nanosensor Technology,* edited by Christine Kranz, 201–31. Cham: Springer. doi:10.1007/5346_2017_10.

Kasibabu, Betha Saineelima B., Stephanie L. D'souza, Sanjay Jha, Rakesh Kumar Singhal, Hirakendu Basu, and Suresh Kumar Kailasa. 2015. "One-Step Synthesis of Fluorescent Carbon Dots for Imaging Bacterial and Fungal Cells." *Analytical Methods* 7 (6). The Royal Society of Chemistry: 2373–78. doi:10.1039/C4AY02737J.

Kelarakis, Antonios. 2014. "From Highly Graphitic to Amorphous Carbon Dots: A Critical Review." *MRS Energy & Sustainability* 1 (July): 1–15. doi:10.1557/mre.2014.7.

Khan, Zubair M. S. H., Shabeena Saifi, Shumaila, Zubair Aslam, Shamshad A. Khan, and M. Zulfequar. 2020. "A Facile One Step Hydrothermal Synthesis of Carbon Quantum Dots for Label -Free Fluorescence Sensing Approach to Detect Picric Acid in Aqueous Solution." *Journal of Photochemistry and Photobiology A: Chemistry* 388: 112201. doi:https://doi.org/10.1016/j.jphotochem.2019.112201.

Kim, Tak H., Hiu Wai Ho, Christopher L. Brown, Sarah L. Cresswell, and Qin Li. 2015. "Amine-Rich Carbon Nanodots as a Fluorescence Probe for Methamphetamine Precursors." *Analytical Methods* 7 (16). The Royal Society of Chemistry: 6869–76. doi:10.1039/C5AY01715G.

Kováčová, Mária, Zoran M. Marković, Petr Humpolíček, Matej Mičušík, Helena Švajdlenková, Angela Kleinová, Martin Danko, et al. 2018. "Carbon Quantum Dots Modified Polyurethane Nanocomposite as Effective Photocatalytic and Antibacterial Agents." *ACS Biomaterials Science & Engineering* 4 (12). American Chemical Society: 3983–93. doi:10.1021/acsbiomaterials.8b00582.

Kumar, Amit, Angshuman Ray Chowdhuri, Dipranjan Laha, Soumen Chandra, Parimal Karmakar, and Sumanta Kumar Sahu. 2016. "One-Pot Synthesis of Carbon Dot-Entrenched Chitosan-Modified Magnetic Nanoparticles for Fluorescence-Based Cu2+ Ion Sensing and Cell Imaging." *RSC Advances* 6 (64). The Royal Society of Chemistry: 58979–87. doi:10.1039/C6RA10382K.

Kuo, Wen-Shuo, Chia-Yuan Chang, Hua-Han Chen, Chih-Li Lilian Hsu, Jiu-Yao Wang, Hui-Fang Kao, Lawrence Chao-Shan Chou, et al. 2016. "Two-Photon Photoexcited Photodynamic Therapy and Contrast Agent with Antimicrobial Graphene Quantum Dots." *ACS Applied Materials & Interfaces* 8 (44). American Chemical Society: 30467–74. doi:10.1021/acsami.6b12014.

Kuo, Wen-Shuo, Yu-Ting Shao, Keng-Shiang Huang, Ting-Mao Chou, and Chih-Hui Yang. 2018. "Antimicrobial Amino-Functionalized Nitrogen-Doped Graphene Quantum Dots for Eliminating Multidrug-Resistant Species in Dual-Modality Photodynamic Therapy and Bioimaging under Two-Photon Excitation." *ACS Applied Materials & Interfaces* 10 (17). American Chemical Society: 14438–46. doi:10.1021/acsami.8b01429.

Lai, Chih-Wei, Yi-Hsuan Hsiao, Yung-Kang Peng, and Pi-Tai Chou. 2012. "Facile Synthesis of Highly Emissive Carbon Dots from Pyrolysis of Glycerol; Gram Scale Production of Carbon Dots/MSiO2 for Cell Imaging and Drug Release." *Journal of Materials Chemistry* 22 (29). The Royal Society of Chemistry: 14403–9. doi:10.1039/C2JM32206D.

Li, Chunmei, Zhaojian Qin, Maonan Wang, Weiwei Liu, Hui Jiang, and Xuemei Wang. 2020. "Manganese Oxide Doped Carbon Dots for Temperature-Responsive Biosensing and Target Bioimaging." *Analytica Chimica Acta* 1104: 125–31. doi:https://doi.org/10.1016/j.aca.2020.01.001.

Li, Hao, Jian Huang, Yuxiang Song, Mengling Zhang, Huibo Wang, Fang Lu, Hui Huang, et al. 2018. "Degradable Carbon Dots with Broad-Spectrum Antibacterial Activity." *ACS Applied Materials & Interfaces* 10 (32). American Chemical Society: 26936–46. doi:10.1021/acsami.8b08832.

Li, Hua, Fang-Qi Shao, Si-Yuan Zou, Qi-Jing Yang, Hong Huang, Jiu-Ju Feng, and Ai-Jun Wang. 2016. "Microwave-Assisted Synthesis of N,P-Doped Carbon Dots for Fluorescent Cell Imaging." *Microchimica Acta* 183 (February). doi:10.1007/s00604-015-1714-2.

Li, Qiqi, Sheng Zhang, Liming Dai, and Liang-shi Li. 2012. "Nitrogen-Doped Colloidal Graphene Quantum Dots and Their Size-Dependent Electrocatalytic Activity for the Oxygen Reduction Reaction." *Journal of the American Chemical Society* 134 (46). American Chemical Society: 18932–35. doi:10.1021/ja309270h.

Li, Shanghao, Daniel Amat, Zhili Peng, Steven Vanni, Scott Raskin, Guillermo De Angulo, Abdelhameed M. Othman, Regina M. Graham, and Roger M. Leblanc. 2016. "Transferrin Conjugated Nontoxic Carbon Dots for Doxorubicin Delivery to Target Pediatric Brain Tumor Cells." *Nanoscale* 8 (37). The Royal Society of Chemistry: 16662–69. doi:10.1039/C6NR05055G.

Li, Shuhua, Yunchao Li, Jun Cao, Jia Zhu, Louzhen Fan, and Xiaohong Li. 2014. "Sulfur-Doped Graphene Quantum Dots as a Novel Fluorescent Probe for Highly Selective and Sensitive Detection of Fe 3+." *Analytical Chemistry* 8 (20): 10201–7.

Li, Xiangyou, Hongqiang Wang, Yoshiki Shimizu, Alexander Pyatenko, Kenji Kawaguchi, and Naoto Koshizaki. 2011. "Preparation of Carbon Quantum Dots with Tunable Photoluminescence by Rapid Laser Passivation in Ordinary Organic Solvents." *Chemical Communications* 47 (3). The Royal Society of Chemistry: 932–34. doi:10.1039/C0CC03552A.

Li, Yu-Jia, Scott G. Harroun, Yu-Chia Su, Chun-Fang Huang, Binesh Unnikrishnan, Han-Jia Lin, Chia-Hua Lin, and Chih-Ching Huang. 2016. "Synthesis of Self-Assembled Spermidine-Carbon Quantum Dots Effective against Multidrug-Resistant Bacteria." *Advanced Healthcare Materials* 5 (19). Wiley: 2545–54. doi:10.1002/adhm.201600297.

Lim, Shi Ying, Wei Shen, and Zhiqiang Gao. 2015. "Carbon Quantum Dots and Their Applications." *Chemical Society Reviews* 44 (1). The Royal Society of Chemistry: 362–81. doi:10.1039/C4CS00269E.

Liu, Changjun, Peng Zhang, Xinyun Zhai, Feng Tian, Wenchen Li, Jianhai Yang, Yuan Liu, Hongbo Wang, Wei Wang, and Wenguang Liu. 2012. "Nano-Carrier for Gene Delivery and Bioimaging Based on Carbon Dots with PEI-Passivation Enhanced Fluorescence." *Biomaterials* 33 (13): 3604–13. doi:https://doi.org/10.1016/j.biomaterials.2012.01.052.

Liu, Haifang, Yuanqiang Sun, Jie Yang, Yalei Hu, Ran Yang, Zhaohui Li, Lingbo Qu, and Yuehe Lin. 2019. "High Performance Fluorescence Biosensing of Cysteine in Human Serum with Superior Specificity Based on Carbon Dots and Cobalt-Derived Recognition." *Sensors and Actuators B: Chemical* 280: 62–68. doi:https://doi.org/10.1016/j.snb.2018.10.029.

Liu, Juanjuan, Yonglei Chen, Weifeng Wang, Jie Feng, Meijuan Liang, Sudai Ma, and Xingguo Chen. 2016. "'Switch-On' Fluorescent Sensing of Ascorbic Acid in Food Samples Based on Carbon Quantum Dots–MnO2 Probe." *Journal of Agricultural and Food Chemistry* 64 (1). American Chemical Society: 371–80. doi:10.1021/acs.jafc.5b05726.

Liu, Junjun, Siyu Lu, Qiuling Tang, Kai Zhang, Weixian Yu, Hongchen Sun, and Bai Yang. 2017. "One-Step Hydrothermal Synthesis of Photoluminescent Carbon Nanodots with Selective Antibacterial Activity against Porphyromonas Gingivalis." *Nanoscale* 9 (21). The Royal Society of Chemistry: 7135–42. doi:10.1039/C7NR02128C.

Liu, Mengli, Yuanhong Xu, Fushuang Niu, J. Justin Gooding, and Jingquan Liu. 2016. "Carbon Quantum Dots Directly Generated from Electrochemical Oxidation of Graphite Electrodes in Alkaline Alcohols and the Applications for Specific Ferric Ion Detection and Cell Imaging." *Analyst* 141 (9). The Royal Society of Chemistry: 2657–64. doi:10.1039/C5AN02231B.

Liu, Ruili, Dongqing Wu, and Xinliang Feng. 2011. "Bottom-Up Fabrication of Photoluminescent Graphene Quantum Dots with Uniform Morphology." *Journal of American Chemical Society* 133 (39): 15221–23.

Liu, Yi, Ning Xiao, Ningqiang Gong, Hao Wang, Xin Shi, Wei Gu, and Ling Ye. 2014. "One-Step Microwave-Assisted Polyol Synthesis of Green Luminescent Carbon Dots as Optical Nanoprobes." *Carbon* 68: 258–64. doi:https://doi.org/10.1016/j.carbon.2013.10.086.

Luo, Tian-Ying, Xi He, Ji Zhang, Ping Chen, Yan-Hong Liu, Hai-Jiao Wang, and Xiao-Qi Yu. 2018. "Photoluminescent F-Doped Carbon Dots Prepared by Ring-Opening Reaction for Gene Delivery and Cell Imaging." *RSC Advances* 8 (11). The Royal Society of Chemistry: 6053–62. doi:10.1039/C7RA13607B.

Ma, Chen'ao, Chaoshun Yin, Yujuan Fan, Xingfa Yang, and Xingping Zhou. 2019. "Highly Efficient Synthesis of N-Doped Carbon Dots with Excellent Stability through Pyrolysis Method." *Journal of Materials Science* 54 (13): 9372–84. doi:10.1007/s10853-019-03585-7.

de Medeiros, Tayline V., John Manioudakis, Farah Noun, Jun-Ray Macairan, Florence Victoria, and Rafik Naccache. 2019. "Microwave-Assisted Synthesis of Carbon Dots and Their Applications." *Journal of Materials Chemistry C* 7 (24). The Royal Society of Chemistry: 7175–95. doi:10.1039/C9TC01640F.

Mehta, Vaibhavkumar N., Shiva Shankaran Chettiar, Jigna R. Bhamore, Suresh Kumar Kailasa, and Ramesh M. Patel. 2017. "Green Synthetic Approach for Synthesis of Fluorescent Carbon Dots for Lisinopril Drug Delivery System and Their Confirmations in the Cells." *Journal of Fluorescence* 27 (1): 111–24. doi:10.1007/s10895-016-1939-4.

Mewada, Ashmi, Sunil Pandey, Mukeshchand Thakur, Dhanashree Jadhav, and Madhuri Sharon. 2014. "Swarming Carbon Dots for Folic Acid Mediated Delivery of Doxorubicin and Biological Imaging." *Journal of Materials Chemistry B* 2 (6). The Royal Society of Chemistry: 698–705. doi:10.1039/C3TB21436B.

Ming, Hai, Zheng Ma, Yang Liu, Keming Pan, Hang Yu, Fang Wang, and Zhenhui Kang. 2012. "Large Scale Electrochemical Synthesis of High Quality Carbon Nanodots and Their Photocatalytic Property." *Dalton Transactions* 41 (31). The Royal Society of Chemistry: 9526–31. doi:10.1039/C2DT30985H.

Mishra, Vijay, Akshay Patil, Sourav Thakur, and Prashant Kesharwani. 2018. "Carbon Dots : Emerging Theranostic Nanoarchitectures." *Drug Discovery Today* 23 (6). Elsevier Ltd: 1219–32. doi:10.1016/j.drudis.2018.01.006.

Mitra, Shouvik, Sourov Chandra, Dipranjan Laha, Prasun Patra, Nitai Debnath, Arindam Pramanik, Panchanan Pramanik, and Arunava Goswami. 2012. "Unique Chemical Grafting of Carbon Nanoparticle on Fabricated ZnO Nanorod: Antibacterial and Bioimaging Property." *Materials Research Bulletin* 47 (3): 586–94. doi:https://doi.org/10.1016/j.materresbull.2011.12.036.

Mueller, Mallory L., Xin Yan, John A. McGuire, and Liang-shi Li. 2010. "Triplet States and Electronic Relaxation in Photoexcited Graphene Quantum Dots." *Nano Letters* 10 (7). American Chemical Society: 2679–82. doi:10.1021/nl101474d.

Namdari, Pooria, Babak Negahdari, and Ali Eatemadi. 2017. "Synthesis, Properties and Biomedical Applications of Carbon-Based Quantum Dots: An Updated Review." *Biomedicine & Pharmacotherapy* 87: 209–22. doi:https://doi.org/10.1016/j.biopha.2016.12.108.

Nguyen, Vanthan, Lihe Yan, Jinhai Si, and Xun Hou. 2015. "Femtosecond Laser-Induced Size Reduction of Carbon Nanodots in Solution: Effect of Laser Fluence, Spot Size, and Irradiation Time." *Journal of Applied Physics* 117 (8). American Institute of Physics: 84304. doi:10.1063/1.4909506.

Pirsaheb, Meghdad, Somayeh Mohammadi, and Abdollah Salimi. 2019. "Current Advances of Carbon Dots Based Biosensors for Tumor Marker Detection, Cancer Cells Analysis and Bioimaging." *TrAC Trends in Analytical Chemistry* 115: 83–99. doi:https://doi.org/10.1016/j.trac.2019.04.003.

Pramanik, Avijit, Stacy Jones, Francisco Pedraza, Aruna Vangara, Carrie Sweet, Mariah S. Williams, Vikram Ruppa-Kasani, Sean Edward Risher, Dhiraj Sardar, and Paresh Chandra Ray. 2017. "Fluorescent, Magnetic Multifunctional Carbon Dots for Selective Separation, Identification, and Eradication of Drug-Resistant Superbugs." *ACS Omega* 2 (2). American Chemical Society: 554–62. doi:10.1021/acsomega.6b00518.

Priyadarshini, Eepsita, Kamla Rawat, Tulika Prasad, and H. B. Bohidar. 2018. "Antifungal Efficacy of Au@ Carbon Dots Nanoconjugates against Opportunistic Fungal Pathogen, Candida Albicans." *Colloids and Surfaces B: Biointerfaces* 163: 355–61. doi:https://doi.org/10.1016/j.colsurfb.2018.01.006.

Qu, Dan, Min Zheng, Peng Du, Yue Zhou, Ligong Zhang, Di Li, Huaqiao Tan, Zhao Zhao, Zhigang Xie, and Zaicheng Sun. 2013. "Highly Luminescent S, N Co-Doped Graphene Quantum Dots with Broad Visible Absorption Bands for Visible Light Photocatalysts." *Nanoscale* 5 (24). The Royal Society of Chemistry: 12272–77. doi:10.1039/C3NR04402E.

Qu, Jia-Huan, Qingyi Wei, and Da-Wen Sun. 2018. "Carbon Dots: Principles and Their Applications in Food Quality and Safety Detection." *Critical Reviews in Food Science and Nutrition* 58(February). Taylor & Francis: 1–10. doi:10.1080/10408398.2018.1437712.

Rao, Longshi, Yong Tang, Zongtao Li, Xinrui Ding, Guanwei Liang, Hanguang Lu, Caiman Yan, Kairui Tang, and Binhai Yu. 2017. "Efficient Synthesis of Highly Fluorescent Carbon Dots by Microreactor Method and Their Application in Fe3+ Ion Detection." *Materials Science and Engineering: C* 81: 213–23. doi:https://doi.org/10.1016/j.msec.2017.07.046.

Ray Chowdhuri, Angshuman, Satyajit Tripathy, Chanchal Haldar, Somenath Roy, and Sumanta Kumar Sahu. 2015. “Single Step Synthesis of Carbon Dot Embedded Chitosan Nanoparticles for Cell Imaging and Hydrophobic Drug Delivery.” *Journal of Materials Chemistry B* 3 (47). The Royal Society of Chemistry: 9122–31. doi:10.1039/C5TB01831E.

Reyes, Delfino, Marco Camacho, Miguel Camacho, Miguel Mayorga, Duncan Weathers, Greg Salamo, Zhiming Wang, and Arup Neogi. 2016. “Laser Ablated Carbon Nanodots for Light Emission.” *Nanoscale Research Letters* 11 (1). Springer US: 424. doi:10.1186/s11671-016-1638-8.

Roy, Prathik, Po-Cheng Chen, Arun Prakash Periasamy, Ya-Na Chen, and Huan-Tsung Chang. 2015. “Photoluminescent Carbon Nanodots: Synthesis, Physicochemical Properties and Analytical Applications.” *Materials Today* 18 (8): 447–58. doi:https://doi.org/10.1016/j.mattod.2015.04.005.

Sarkar, Tamal, H. B. Bohidar, and Pratima R. Solanki. 2018. “Carbon Dots-Modified Chitosan Based Electrochemical Biosensing Platform for Detection of Vitamin D.” *International Journal of Biological Macromolecules* 109: 687–97. doi:https://doi.org/10.1016/j.ijbiomac.2017.12.122.

Sharma, Anirudh, and Joydeep Das. 2019. “Small Molecules Derived Carbon Dots: Synthesis and Applications in Sensing, Catalysis, Imaging, and Biomedicine.” *Journal of Nanobiotechnology* 17 (1). BioMed Central: 92. doi:10.1186/s12951-019-0525-8.

Shu, Yang, Jun Lu, Quan-Xing Mao, Ru-Sheng Song, Xue-Ying Wang, Xu-Wei Chen, and Jian-Hua Wang. 2017. “Ionic Liquid Mediated Organophilic Carbon Dots for Drug Delivery and Bioimaging.” *Carbon* 114: 324–33. doi:https://doi.org/10.1016/j.carbon.2016.12.038.

Singh, Seema, Anshul Mishra, Rina Kumari, Kislay K. Sinha, Manoj K. Singh, and Prolay Das. 2017. “Carbon Dots Assisted Formation of DNA Hydrogel for Sustained Release of Drug.” *Carbon* 114: 169–76. doi:https://doi.org/10.1016/j.carbon.2016.12.020.

Sonthanasamy, Regina Sisika A., Azwan Mat Lazim, Siti Nur Syazni Mohd Zuki, Doris Huai Xia Quay, and Ling Ling Tan. 2020. “Starch-Based C-Dots from Natural Gadong Tuber as PH Fluorescence Label for Optical Biosensing of Arginine.” *Optics & Laser Technology* 130: 106345. doi:https://doi.org/10.1016/j.optlastec.2020.106345.

Sun, Ya-Ping, Bing Zhou, Yi Lin, Wei Wang, KA Shiral Fernando, Pankaj Pathak, Mohammed Jaouad Meziani, et al. 2006. “Quantum-Sized Carbon Dots for Bright and Colorful Photoluminescence.” *Journal of the American Chemical Society* 128 (24). American Chemical Society: 7756–57. doi:10.1021/ja062677d.

Tang, Jing, Biao Kong, Hao Wu, Ming Xu, Yongcheng Wang, Yanli Wang, Dongyuan Zhao, and Gengfeng Zheng. 2013. “Carbon Nanodots Featuring Efficient FRET for Real-Time Monitoring of Drug Delivery and Two-Photon Imaging.” *Advanced Materials (Deerfield Beach, Fla.)* 25 (45). Laboratory of Advanced Materials Department of Chemistry, Fudan University, Shanghai, China: 6569–74. doi:10.1002/adma.201303124.

Tang, Libin, Rongbin Ji, Xueming Li, Gongxun Bai, Chao Ping Liu, Jianhua Hao, Jingyu Lin, et al. 2014. “Deep Ultraviolet to Near-Infrared Emission and Photoresponse in Layered N-Doped Graphene Quantum Dots.” *ACS Nano* 8 (6). American Chemical Society: 6312–20. doi:10.1021/nn501796r.

Thakur, Mukeshchand, Sunil Pandey, Ashmi Mewada, Vaibhav Patil, Monika Khade, Ekta Goshi, and Madhuri Sharon. 2014. “Antibiotic Conjugated Fluorescent Carbon Dots as a Theranostic Agent for Controlled Drug Release, Bioimaging, and Enhanced Antimicrobial Activity.” Edited by Roberta Cavalli. *Journal of Drug Delivery* 2014. Hindawi Publishing Corporation: 282193. doi:10.1155/2014/282193.

Tuerhong, Mhetaer, Yang XU, and Xue-Bo YIN. 2017. "Review on Carbon Dots and Their Applications." *Chinese Journal of Analytical Chemistry* 45 (1): 139–50. doi:https://doi.org/10.1016/S1872-2040(16)60990-8.

Wang, Beibei, Shujun Wang, Yanfang Wang, Yan Lv, Hao Wu, Xiaojun Ma, and Mingqian Tan. 2016. "Highly Fluorescent Carbon Dots for Visible Sensing of Doxorubicin Release Based on Efficient Nanosurface Energy Transfer." *Biotechnology Letters* 38 (1): 191–201. doi:10.1007/s10529-015-1965-3.

Wang, Bin, Yanfen Chen, Yuanya Wu, Bo Weng, Yingshuai Liu, Zhisong Lu, Chang Ming Li, and Cong Yu. 2016. "Aptamer Induced Assembly of Fluorescent Nitrogen-Doped Carbon Dots on Gold Nanoparticles for Sensitive Detection of AFB1." *Biosensors and Bioelectronics* 78: 23–30. doi:https://doi.org/10.1016/j.bios.2015.11.015.

Wang, Duo, Bixia Lin, Yujuan Cao, Manli Guo, and Ying Yu. 2016. "A Highly Selective and Sensitive Fluorescence Detection Method of Glyphosate Based on an Immune Reaction Strategy of Carbon Dot Labeled Antibody and Antigen Magnetic Beads." *Journal of Agricultural and Food Chemistry* 64 (30). American Chemical Society: 6042–50. doi:10.1021/acs.jafc.6b01088.

Wang, Kan, Zhongcai Gao, Guo Gao, Yan Wo, Yuxia Wang, Guangxia Shen, and Daxiang Cui. 2013. "Systematic Safety Evaluation on Photoluminescent Carbon Dots." *Nanoscale Research Letters* 8 (1): 122. doi:10.1186/1556-276X-8-122.

Wang, Qing, Chunlei Zhang, Guangxia Shen, Huiyang Liu, Hualin Fu, and Daxiang Cui. 2014. "Fluorescent Carbon Dots as an Efficient SiRNA Nanocarrier for Its Interference Therapy in Gastric Cancer Cells." *Journal of Nanobiotechnology* 12 (1): 58. doi:10.1186/s12951-014-0058-0.

Wang, Qinlong, Xiaoxiao Huang, Yijuan Long, Xiliang Wang, Haijie Zhang, Rui Zhu, Liping Liang, Ping Teng, and Huzhi Zheng. 2013. "Hollow Luminescent Carbon Dots for Drug Delivery." *Carbon* 59: 192–99. doi:https://doi.org/10.1016/j.carbon.2013.03.009.

Wang, Renjie, Yi Xu, Tao Zhang, and Yan Jiang. 2015. "Rapid and Sensitive Detection of Salmonella Typhimurium Using Aptamer-Conjugated Carbon Dots as Fluorescence Probe." *Analytical Methods* 7 (5). The Royal Society of Chemistry: 1701–6. doi:10.1039/C4AY02880E.

Wang, Weiping, Ya-Chun Lu, Hong Huang, Jiu-Ju Feng, Jian-Rong Chen, and Ai-Jun Wang. 2014. "Facile Synthesis of Water-Soluble and Biocompatible Fluorescent Nitrogen-Doped Carbon Dots for Cell Imaging." *Analyst* 139 (7). The Royal Society of Chemistry: 1692–96. doi:10.1039/C3AN02098C.

Wang, Xiao, Yongqiang Feng, Peipei Dong, and Jianfeng Huang. 2019. "A Mini Review on Carbon Quantum Dots: Preparation, Properties, and Electrocatalytic Application." *Frontiers in Chemistry*. https://www.frontiersin.org/article/10.3389/fchem.2019.00671.

Wu, Yu-Fen, Hsi-Chin Wu, Chen-Hsiang Kuan, Chun-Jui Lin, Li-Wen Wang, Chien-Wen Chang, and Tzu-Wei Wang. 2016. "Multi-Functionalized Carbon Dots as Theranostic Nanoagent for Gene Delivery in Lung Cancer Therapy." *Scientific Reports* 6 (February). Nature Publishing Group: 21170. doi:10.1038/srep21170.

Wu, Zhu Lian, Ze Xi Liu, and Yun Huan Yuan. 2017. "Carbon Dots: Materials, Synthesis, Properties and Approaches to Long-Wavelength and Multicolor Emission." *Journal of Materials Chemistry B* 5 (21). The Royal Society of Chemistry: 3794–3809. doi:10.1039/C7TB00363C.

Wu, Zhu Lian, Pu Zhang, Ming Xuan Gao, Chun Fang Liu, Wei Wang, Fei Leng, and Cheng Zhi Huang. 2013. "One-Pot Hydrothermal Synthesis of Highly Luminescent Nitrogen-Doped Amphoteric Carbon Dots for Bioimaging from Bombyx Mori Silk – Natural Proteins." *Journal of Materials Chemistry B* 1 (22). The Royal Society of Chemistry: 2868–73. doi:10.1039/C3TB20418A.

Xiang, Yiming, Congyang Mao, Xiangmei Liu, Zhenduo Cui, Doudou Jing, Xianjin Yang, Yanqin Liang, et al. 2019. "Rapid and Superior Bacteria Killing of Carbon Quantum Dots/ZnO Decorated Injectable Folic Acid-Conjugated PDA Hydrogel through Dual-Light Triggered ROS and Membrane Permeability." *Small* 15 (22). Wiley: 1900322. doi:10.1002/smll.201900322.

Xiao, J., P. Liu, C. X. Wang, and G. W. Yang. 2017. "External Field-Assisted Laser Ablation in Liquid: An Efficient Strategy for Nanocrystal Synthesis and Nanostructure Assembly." *Progress in Materials Science* 87: 140–220. doi:https://doi.org/10.1016/j.pmatsci.2017.02.004.

Xu, Hua, Xiupei Yang, Gu Li, Chuan Zhao, and Xiangjun Liao. 2015. "Green Synthesis of Fluorescent Carbon Dots for Selective Detection of Tartrazine in Food Samples." *Journal of Agricultural and Food Chemistry* 63 (30). American Chemical Society: 6707–14. doi:10.1021/acs.jafc.5b02319.

Xu, Xiaoyou, Robert Ray, Yunlong Gu, Harry J. Ploehn, Latha Gearheart, Kyle Raker, and Walter A. Scrivens. 2004. "Electrophoretic Analysis and Purification of Fluorescent Single-Walled Carbon Nanotube Fragments." *Journal of the American Chemical Society* 126 (40). American Chemical Society: 12736–37. doi:10.1021/ja040082h.

Yan, Yayuan, Weicong Kuang, Liujun Shi, Xiaoli Ye, Yunhua Yang, Xiaobao Xie, Qingshan Shi, and Shaozao Tan. 2019. "Carbon Quantum Dot-Decorated TiO2 for Fast and Sustainable Antibacterial Properties under Visible-Light." *Journal of Alloys and Compounds* 777: 234–43. doi:https://doi.org/10.1016/j.jallcom.2018.10.191.

Yang, Chuanxu, Rasmus Peter Thomsen, Ryosuke Ogaki, Jørgen Kjems, and Boon M. Teo. 2015. "Ultrastable Green Fluorescence Carbon Dots with a High Quantum Yield for Bioimaging and Use as Theranostic Carriers." *Journal of Materials Chemistry B* 3 (22). The Royal Society of Chemistry: 4577–84. doi:10.1039/C5TB00467E.

Yang, Kuncheng, Shanshan Wang, Yingyi Wang, Hong Miao, and Xiaoming Yang. 2017. "Dual-Channel Probe of Carbon Dots Cooperating with Gold Nanoclusters Employed for Assaying Multiple Targets." *Biosensors and Bioelectronics* 91: 566–73. doi:https://doi.org/10.1016/j.bios.2017.01.014.

Yang, Lei, Zheran Wang, Ju Wang, Weihua Jiang, Xuewei Jiang, Zhaoshi Bai, Yunpeng He, Jianqi Jiang, Dongkai Wang, and Li Yang. 2016. "Doxorubicin Conjugated Functionalizable Carbon Dots for Nucleus Targeted Delivery and Enhanced Therapeutic Efficacy." *Nanoscale* 8 (12). The Royal Society of Chemistry: 6801–9. doi:10.1039/C6NR00247A.

Yang, Sheng-Tao, Xin Wang, Haifang Wang, Fushen Lu, Pengju G. Luo, Li Cao, Mohammed J. Meziani, et al. 2009. "Carbon Dots as Nontoxic and High-Performance Fluorescence Imaging Agents." *The Journal of Physical Chemistry C* 113 (42). American Chemical Society: 18110–14. doi:10.1021/jp9085969.

Yang, Siwei, Jing Sun, Xiubing Li, Wei Zhou, Zhongyang Wang, Pen He, Guqiao Ding, Xiaoming Xie, Zhenhui Kang, and Mianheng Jiang. 2014. "Large-Scale Fabrication of Heavy Doped Carbon Quantum Dots with Tunable-Photoluminescence and Sensitive Fluorescence Detection." *Journal of Materials Chemistry A* 2 (23). The Royal Society of Chemistry: 8660–67. doi:10.1039/C4TA00860J.

Yang, Xiaoming, Yawen Luo, Shanshan Zhu, Yuanjiao Feng, Yan Zhuo, and Yao Dou. 2014. "One-Pot Synthesis of High Fluorescent Carbon Nanoparticles and Their Applications as Probes for Detection of Tetracyclines." *Biosensors and Bioelectronics* 56: 6–11. doi:https://doi.org/10.1016/j.bios.2013.12.064.

Yang, Zheng-Chun, Xu Li, and John Wang. 2011a. "Intrinsically Fluorescent Nitrogen-Containing Carbon Nanoparticles Synthesized by a Hydrothermal Process." *Carbon* 49 (15): 5207–12. doi:https://doi.org/10.1016/j.carbon.2011.07.038.

Yang, Zheng-Chun, Miao Wang, Anna Marie Yong, Siew Yee Wong, Xin-Hai Zhang, Happy Tan, Alex Yuangchi Chang, Xu Li, and John Wang. 2011b. "Intrinsically Fluorescent Carbon Dots with Tunable Emission Derived from Hydrothermal Treatment of Glucose in the Presence of Monopotassium Phosphate." *Chemical Communications* 47 (42). The Royal Society of Chemistry: 11615–17. doi:10.1039/C1CC14860E.

Yu, Ying, Chunguang Li, Cailing Chen, He Huang, Chen Liang, Yue Lou, Xiao-Bo Chen, Zhan Shi, and Shouhua Feng. 2019. "Saccharomyces-Derived Carbon Dots for Biosensing PH and Vitamin B 12." *Talanta* 195: 117–26. doi:https://doi.org/10.1016/j.talanta.2018.11.010.

Yuan, Fanglong, Ling Ding, Yunchao Li, Xiaohong Li, Louzhen Fan, Shixin Zhou, Decai Fang, and Shihe Yang. 2015. "Multicolor Fluorescent Graphene Quantum Dots Colorimetrically Responsive to All-PH and a Wide Temperature Range." *Nanoscale* 7 (27). The Royal Society of Chemistry: 11727–33. doi:10.1039/C5NR02007G.

Yuan, Fanglong, Shuhua Li, Zetan Fan, Xiangyue Meng, Louzhen Fan, and Shihe Yang. 2016. "Shining Carbon Dots: Synthesis and Biomedical and Optoelectronic Applications." *Nano Today* 11 (5). Elsevier Ltd: 565–86. doi:10.1016/j.nantod.2016.08.006.

Zeng, Qinghui, Dan Shao, Xu He, Zhongyuan Ren, Wenyu Ji, Chongxin Shan, Songnan Qu, Jing Li, Li Chen, and Qin Li. 2016. "Carbon Dots as a Trackable Drug Delivery Carrier for Localized Cancer Therapy in Vivo." *Journal of Materials Chemistry B* 4 (30). The Royal Society of Chemistry: 5119–26. doi:10.1039/C6TB01259K.

Zeng, Ya-Wen, De-Kun Ma, Wei Wang, Jing-Jing Chen, Lin Zhou, Yi-Zhou Zheng, Kang Yu, and Shao-Ming Huang. 2015. "N, S Co-Doped Carbon Dots with Orange Luminescence Synthesized through Polymerization and Carbonization Reaction of Amino Acids." *Applied Surface Science* 342: 136–43. doi:https://doi.org/10.1016/j.apsusc.2015.03.029.

Zhang, Bing, Chun-yan Liu, and Yun Liu. 2010. "A Novel One-Step Approach to Synthesize Fluorescent Carbon Nanoparticles." *European Journal of Inorganic Chemistry* 2010 (28). John: 4411–14. doi:https://doi.org/10.1002/ejic.201000622.

Zhang, Jia, and Shu-Hong Yu. 2016. "Carbon Dots: Large-Scale Synthesis, Sensing and Bioimaging." *Materials Today* 19 (7): 382–93. doi:https://doi.org/10.1016/j.mattod.2015.11.008.

Zhang, M., X. Zhao, Z. Fang, Y. Niu, J. Lou, Y. Wu, S. Zou, S. Xia, M. Sun, and F. Du. 2017. "Fabrication of HA/PEI-Functionalized Carbon Dots for Tumor Targeting, Intracellular Imaging and Gene Delivery." *RSC Advances* 7 (6). The Royal Society of Chemistry: 3369–75. doi:10.1039/C6RA26048A.

Zhang, Qinghong, Xiaofeng Sun, Hong Ruan, Keyang Yin, and Hongguang Li. 2017. "Production of Yellow-Emitting Carbon Quantum Dots from Fullerene Carbon Soot." *Science China Materials* 60 (2): 141–50. doi:10.1007/s40843-016-5160-9.

Zhao, Yongqiang, Xuguang Liu, Yongzhen Yang, Litao Kang, Zhi Yang, Weifeng Liu, and Lin Chen. 2015. "Carbon Dots: From Intense Absorption in Visible Range to Excitation-Independent and Excitation-Dependent Photoluminescence." *Fullerenes, Nanotubes and Carbon Nanostructures* 23 (11). Taylor & Francis: 922–29. doi:10.1080/1536383X.2015.1018413.

Zheng, Min, Shi Liu, Jing Li, Dan Qu, Haifeng Zhao, Xingang Guan, Xiuli Hu, Zhigang Xie, Xiabin Jing, and Zaicheng Sun. 2014. "Integrating Oxaliplatin with Highly Luminescent Carbon Dots: An Unprecedented Theranostic Agent for Personalized Medicine." *Advanced Materials* 26 (21). John: 3554–60. doi:https://doi.org/10.1002/adma.201306192.

Zhou, Jigang, Christina Booker, Ruying Li, Xingtai Zhou, Tsun-Kong Sham, Xueliang Sun, and Zhifeng Ding. 2007. "An Electrochemical Avenue to Blue Luminescent Nanocrystals from Multiwalled Carbon Nanotubes (MWCNTs)." *Journal of the American Chemical Society* 129 (4). American Chemical Society: 744–45. doi:10.1021/ja0669070.

Zhou, Langfeng, Meng Qiao, Lei Zhang, Lu Sun, Yang Zhang, and Weiwei Liu. 2019. "Green and Efficient Synthesis of Carbon Quantum Dots and Their Luminescent Properties." *Journal of Luminescence* 206: 158–63. doi:https://doi.org/10.1016/j.jlumin.2018.10.057.

Zhou, Li, Zhenhua Li, Zhen Liu, Jinsong Ren, and Xiaogang Qu. 2013. "Luminescent Carbon Dot-Gated Nanovehicles for PH-Triggered Intracellular Controlled Release and Imaging." *Langmuir* 29 (21). American Chemical Society: 6396–6403. doi:10.1021/la400479n.

4 Carbon Nanotubes
Synthesis, Properties, and Modifications

Deepshikha Rathore and Umesh K. Dwivedi

CONTENTS

4.1 INTRODUCTION

In 1991, Iijima discovered carbon nanotubes (CNTs), which have generated significant attention following extensive research into the synthesis and properties of a cylindrical carbon matrix in the nano range. A CNT consists of a honeycomb lattice folded into a cylinder (Iijima 1991). The multi-walled carbon nanotubes (MWCNTs) usually possess 2 to 50 successive cylindrical shells with inherent sp^2 hybridized C=C bonds. Typically, they are several tens of nm in diameter, with length in the μm range. They originate on the tip of the cathode in DC arc deposition, as described in the synthesis section (Section 4.2; Kiang et al. 1995).

The diameter of CNTs is much smaller than the most innovative semiconductor devices available on the market. Accordingly, the accessibility of CNTs may have a significant influence on semiconductor technology due to their small diameter and versatile electronic and transport properties. These remarkable, physical properties of CNTs are generated because of their *chirality* (helical geometrical structure). Specifically, the complete electronic structure of single-walled carbon

DOI: 10.1201/9781003110781-4

nanotubes (SWCNTs), which are either semiconducting or metallic, is governed by their *chirality* and *diameter*. Therefore, many potential semiconductor devices at the nanoscale can be fabricated using CNTs. Additionally, it has been observed by many researchers that, by changing the diameter of semiconducting CNTs, the energy gap can be changed continuously from 1 eV to 0 eV. Consequently, with the help of the geometrical structure of carbon atoms, with sp^2 inherent hybridization of C=C bond, many desirable semiconducting properties can be defined precisely (Saito et al. 1998).

This chapter describes synthesis techniques and various potential properties of CNTs. Three comparatively efficient approaches to synthesizing SWCNTs have been recognized: chemical vapour deposition, electric arc, and laser deposition. The purification of grown CNTs will also be discussed with all physical properties, namely electrical, transport, mechanical, thermal, vibrational, and elastic properties. The modification technique for CNTs will be described at the end of the chapter, in Section 4.5.

4.2 SYNTHESIS OF CNTs

Three main fabrication techniques have been used to the current day to produce single= and multi-walled carbon nanotubes. These processes will be described here in Section 4.2.

4.2.1 Chemical Vapor Deposition

In 1993, Endo et al. first produced effective MWCNTs, using chemical vapour deposition (CVD) (Endo et al. 1993, 1995). At Rice University, in 1996, Dai, an active member of Smalley's group, developed CO-based CVDs (chemical vapor deposition) to synthesize SWCNTs (O'Connell 2006). The CVD technique incorporates a wide range of synthetic methodologies. CVDs can produce gram-quantities of CNTs as bulk-formed or individually arranged SWCNTs on SiO_2 substrate for electronic applications. CVDs can also yield vertically aligned MWCNTs for the fabrication of high-performance field emitters. Moreover, CVDs can create SWCNTs with advanced atomic quality and greater formation rate than existing synthesis techniques. Recently, various CNT production techniques have been developed, based on CVD.

The synthesis of carbon nanotubes by chemical evapour deposition involves Fe particles as a catalyst with a low benzene gas pressure and the furnace temperature held at 1100°C (Endo et al. 1993, 1995). The fabrication of CNTs is described in Figure 4.1. To fabricate CNTs, numerous kinds of catalysts, catalyst supports, and hydrocarbon components are used frequently by several research groups around the world. The advantage of this technique is that an enormous amount of CNTs can be produced for commercialization under optimum conditions. Directly produced CNTs exhibit decreased crystallinity. The crystallinity can be enhanced in the presence of argon after heat treatment from 2500 to 3000°C.

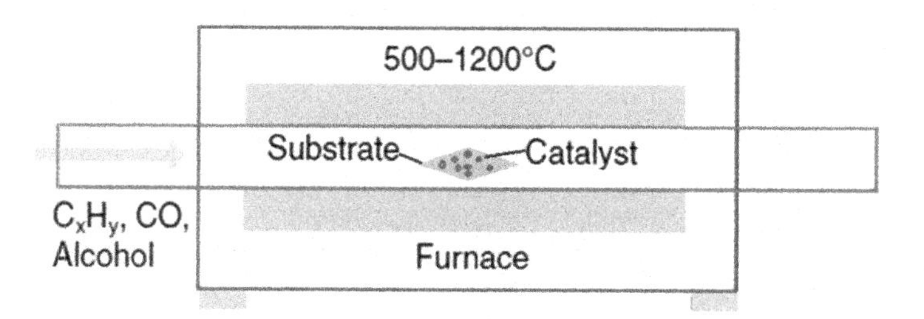

FIGURE 4.1 Schematic of CVD technique to synthesize CNTs.

4.2.2 Electric Arc Method

The electric arc offers a traditional, modest tool for creating a plasma of carbon atoms after vapourization at a temperature greater than 3000°C (Ebbesen and Ajayan 1992, Ebessen et al. 1993). It is also known as the carbon arc technique. With the help of this technique, both single-walled CNTs with ropes and multi-walled CNTs can be synthesized (Journet et al. 1997). To functionalize the electric arc method, the operating conditions comprise (1) graphite electrodes of 5–20 mm diameter, (2) 1 mm separation between electrodes, (3) voltage of 20–25 V across the graphite electrodes, (4) 50–120 A dc electric current flowing between the graphite electrodes, and (5) a flow rate of He maintained at ~5–15ml/s at 500 torr pressure to cool the system.

In this technique, the positive electrode acts as the anode, and the negative electrode works as the cathode. During the synthesis of CNTs, the length and thickness of the anode decreases, and carbon atoms start to deposit on the cathode, as shown in Figure 4.2. No catalyst is needed to synthesize MWCNTs, and they are found in bundles in the interior part of the negative cathode electrode where the temperature is at its maximum value (2500–3000°C). These bundles of CNTs are unevenly arranged in the direction of electric current flow (see Figure 4.2) (Ebbesen and Ajayan 1992, Ebessen 1994). Near synthesized CNTs, a rigid shell is also produced, containing fullerene, nanoparticles, and amorphous carbon atoms (Ajayan and Lijima 1992, Dravid et al. 1993, Iijima 1993). To increase the yield of CNTs, a suitable cooling process is required in the growth chamber. Catalysts play a significant role in the process of fabricating isolated SWCNTs with ropes. Rare earths (Y and Gd), transition metals (Co, Ni, and Fe) and miscellaneous combinations (Fe/Ni, Co/Ni and Co/Pt) have been utilized as catalysts up to the present by researchers.

Some critical parameters, such as the temperature of the growth chamber, synthesis conditions, and the presence of a catalyst, can control the diameter and chirality of the SWCNTs. In this process, the average diameter of CNTs is generally less (~50 nm), and the distribution of diameters is typically narrow. It is remarkable to realize that the minimum diameter of SWCNTs is ~0.7 nm, comparable to the diameter of fullerene (C_{60}), which follows the isolated pentagon rule (Smalley 1992). With the help of the electric arc technique, large g-scale bundles of SWCNTs have been successfully generated (Journet et al. 1997). On the collarette surrounding the cathode,

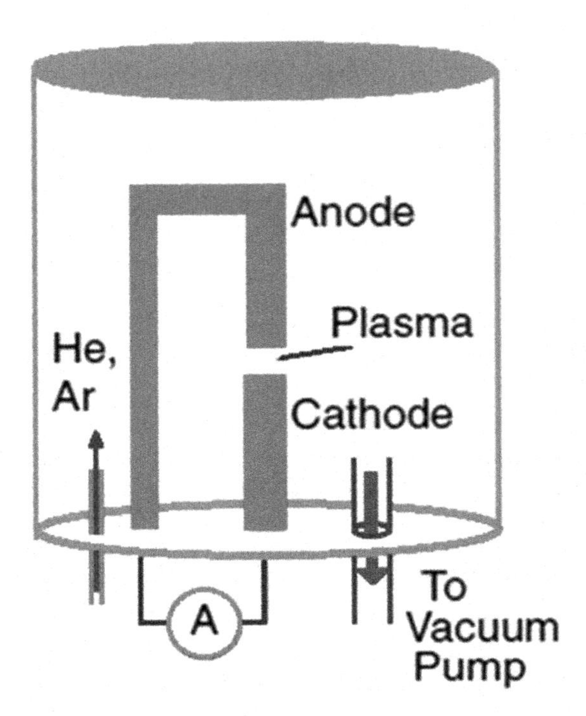

FIGURE 4.2 Schematic of electric arc method.

the maximum amount of SWCNTs originates, almost 20 % of the excessive carbon being transformed into SWCNTs on the collarette. Subsequently, the length of CNTs is usually on the order of 1 μm with a growth period projected to be less than 0.1 seconds, whereas a single circumferential carbon ring grows in 10^{-5} seconds. This is so much faster than the time required for the direct investigation of the graphite arc electrodes by transmission electron microscopy (TEM).

On the other hand, this growth period is much longer than the phonon frequency of 10^{15} Hz. Therefore, the growth of CNTs follows under controlled conditions with maintained electric current and arc voltage over a limited range of He or Ar gas pressure. Hence, cleansing and purification are essential for dispersing pure SWCNTs from the mixture with the above impurities. In this method, due to the cobalt or iron catalysts involved in SWCNTs synthesis, carbide nanoparticles surrounding the graphite cathode electrode and clusters of metal catalyst captured inside the graphite layers are also fabricated (Iijima and Ichihashi 1993, Bethune et al. 1993). Those impurities usually deposit inside the central area of the anode electrode, whereas carbon nanoparticles, in the form of capsules, deposit on the cathode electrode (Saito et al. 1993). Catalyst metal particles, like transition metals (Fe, Co, Ni, Cu and Au), transition metal carbides, and rare earth metal carbides, are confined inside carbon nanocapsules. These carbon nanocapsules consist of multi-walled polyhedra of graphene sheets with interlayer separations of 0.34 nm.

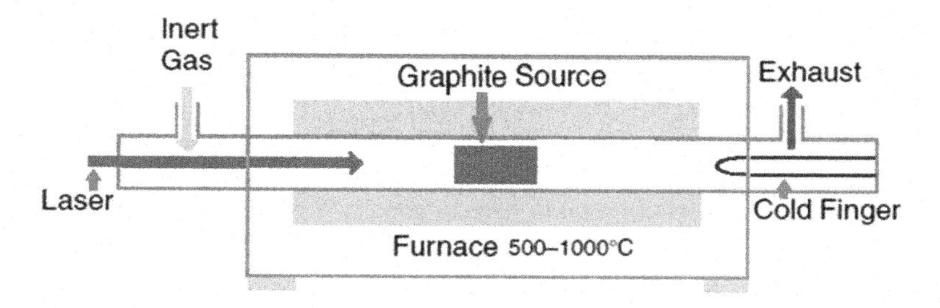

FIGURE 4.3 Schematic of laser deposition technique.

4.2.3 Laser Deposition

The laser deposition technique has been employed to produce bundles of SWCNTs of narrow diameter distribution, using a graphite target. The maximum conversion of graphite into SWCNTs, using laser deposition, is reported to be greater than 70%–90% by condensation of carbon vapour in the heated flow tube, operating at 1200°C (Thess et al. 1996). The Co-Ni/graphite composites are used as a target, containing 3% or 1.2 atoms of Co-Ni alloy and 97% or 98.8 atoms of graphite. There are two successive pulses of laser incident on the graphite target and evaporate it to produce carbon atoms. Some impurities of transition metals (Co and Ni) are also found in the synthesis of SWCNTs. The synthesized SWCNTs flow, with the help of argon gas, from a high-temperature region to a water-cooled Cu collector, which is placed outside the furnace (Thess et al. 1996, Yudasaka et al. 1997) (Figure 4.3).

4.3 PURIFICATION

Researchers have reported that, during the synthesis of CNTs, some other impurities like carbon nanoparticles, amorphous carbon, and catalyst metal particles are also obtained. The removal of these impurities from CNTs is known as purification, although the separation of CNTs according to chirality and diameter does not come under the heading of purification. CNT separation is quite straightforward technologically currently, but removing metal impurities from CNTs is a difficult process, with very limited success being achieved with the help of three traditional techniques: (1) liquid phase, (2) gas phase, and (3) intercalation (Ebbesen 1996). Other standard methods have also been utilized for purification: centrifugation, filtration, and chromatography, but very effective results could not be found after removing unwanted entities. Heat treatment favourably reduces the quantity of carbon-related disorders, but, after heat treatment, the diameter of CNTs expands due to an increase in the epitaxial carbon layers in the vapour phase. With the help of the oxidation and oxygen burning process, the amorphous carbon and other nanoparticles can be removed from CNTs using the gas phase technique (Ajayan et al. 1993, Tsang et al. 1993, Ebessen 1994). The oxidation process is thermally stimulated by the activation energy of 225 kJ/mol in the presence of air (Ajayan et al. 1993). In MWCNTs, removing metal impurities from successive layers is extremely slow because oxygen is trapped much more strongly in the graphene matrix than are other amorphous carbon nanoparticles within pentagonal structures (Tsang et al. 1993, Ajayan et al. 1993). In this process, some CNTs are also burnt off. In the gas phase purification technique, MWCNTs are mainly produced 20–200 Å in diameter and 10 nm to 1 µm in length. The smaller diameter CNTs are used to oxidize nanoparticles. The liquid phase method employs potassium permanganate ($KMnO_4$) treatment to remove unwanted carbon and other nanoparticles and achieves greater yield with a shorter CNT length than the gas phase removal method. Furthermore, in the intercalation method, the CNTs are treated with $CuC1_2$-KCl, and consequent chemical removal of the intercalated species is accomplished (Ikazaki et al. 1994). Ultimately, some carbon nanoparticles, other impurities, and fullerenes are still present in purified CNTs (Rao et al. 1997).

4.4 PROPERTIES

CNTs exhibit many versatile physical properties because of their inherent sp^2 hybridized C=C bond hexagonal helical geometry. These structure-dependent properties can be adjusted by varying the diameter and chirality, which are widely used in current technology for potential applications. All such properties will be described in detail in this section (Section 4.4).

4.4.1 Electrical and Transport Properties

Depending upon their chirality and diameter, CNTs exhibit semiconducting and metallic behaviour with interesting electrical properties. *Chirality* refers to how the graphene sheets are rolled, corresponding to the direction of the translation vector in the axis of CNTs. In the synthesis of CNTs, the mixture of one-third metallic and two-thirds semiconducting properties is obtained. From the metallic CNTs, the mixture of CNTs usually exhibits an *armchair* structure, whereas semiconducting CNTs attain a *chiral* structure. To investigate the electronic structure of the CNTs, scanning electron microscopy (SEM) has been performed. In this experiment, the position of the SEM tip is placed above the CNT. Then, a voltage is applied between the tip and the CNT, and the tunnelling current is measured. The conductance obtained directly measures the local electronic density of states, which describes how close the energy bands are (Poole Jr. and Owens 2003). The experiments also observed that, in response to increasing the diameter of the chiral semiconducting CNTs, the band gap decreases.

At low temperatures, it is well known that a SWCNT acts as a quantum wire in which the electrons can move without being dispersed through scattering centres (Saito et al. 1998). When SWCNTs possess semiconducting properties, electron transport has been reported to occur at a millikelvin temperature on distinct SWCNTs lying along two metal electrodes. For this purpose, a field-effect transistor was fabricated. Over its source and drain electrodes, SWCNTs were deposited, as illustrated in Figure 4.4. As a voltage is applied on its third electrode (gate), a step-like function is observed in the current–voltage characteristics. This step-like feature

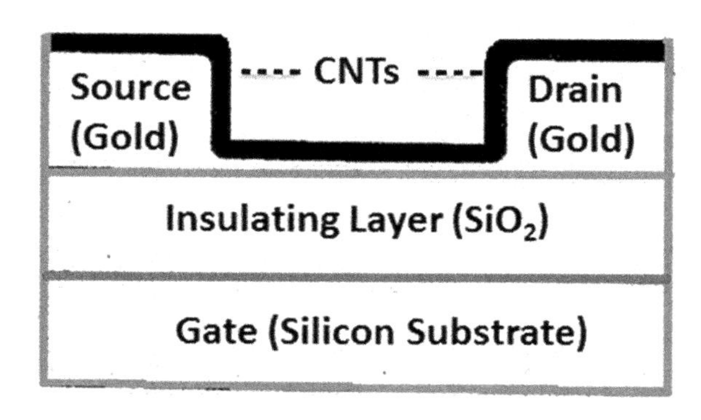

FIGURE 4.4 Schematic of field effect transistor designed by CNTs.

arises due to single-electron transportation through CNTs and the generation of resonant tunnelling *via* single molecular orbitals. When the capacitance of the CNT is significantly less, single-electron tunnelling happens. A single electron is needed to increase an electrostatic charging energy to the thermal energy K_BT. At low voltage, electron transport is blocked, an effect known as a *Coulomb blockade*. Electrons can be added one by one to the CNT on increasing the gate voltage. Electron transport takes place due to electron tunnelling through discrete energy states present in CNTs. The addition of single molecular orbitals initiates the increase in current at each step. From this phenomenon, it can be analyzed that the electrons are not localized at a defined place, but mobile in an extended way in entire CNTs. Usually, the presence of such effects may cause localization of electrons in a one-dimensional system. But in a CNT, the wave function has a doughnut shape, so that the defect will be averaged over the entire circumference of the CNT. Hence, the electron will not be localized at a specific place in the CNT (Poole and Owens 2003).

The conductivity of CNTs is seen to be very high when they exhibit metallic properties. They can carry a current of 10^9 A/cm^2. It is well known that copper (Cu) wire cannot flow at 10^9 A/cm^2 because of resistive heating, which can melt the Cu wire itself. Due to the presence of very few defects as scattering centre, the electrons cannot scatter. Hence, electrical conductivity increases with decreasing resistivity in CNTs. The high current values do not heat the CNTs as occurs when in Cu wire, because they hold very high thermal conductivity.

On the other hand, CNTs exhibit a magnetoresistance phenomenon at low temperature, in which resistance of the CNTs changes when a DC magnetic field is applied. The experiment has been performed by Saito et al. (1998) to show the change in resistance as a function of the magnetic field. They reported that the magnetoresistance effect was negative because they observed a reduction in resistance in response to an increasing DC magnetic field. According to the researchers, conduction electrons apply the DC magnetic field to CNTs to achieve new energy states with their spiraling motion. These energy states are very near to the uppermost occupied energy states known as Landau energy states. Therefore, more energy states occupy the electrons and increase their energy with more conducting CNTs (Poole and Owens 2003).

4.4.2 Mechanical Properties

CNTs are identified as the most robust nanocarbon material. To understand their mechanical properties, stress, strain, Young's modulus etc. need to be defined. It is well known that If weight W is connected by one end of the rod and the other end is nailed to the ceiling of the room, then stress S acts on the rod as weight per unit cross-sectional area A of the rod:

$$S = \frac{W}{A} \tag{1}$$

The strain ε is generated along the rod as the quantity of elongation ΔL of the rod per unit length L of the rod.

$$\varepsilon = \frac{\Delta L}{L} \tag{2}$$

According to Hooke's law, the elongation ΔL of the rod is directly proportional to the W weight connected with the rod. In general, stress S is directly proportional to strain ε:

$$S = Y\varepsilon \tag{3}$$

where Young's modulus Y is a proportionality constant.

$$Y = \frac{WL}{\Delta LA} \tag{4}$$

It is the elastic and flexible property characteristic of the material. It can be concluded from equation (4) of Young's modulus that the greater the value of Young's modulus, the lesser is the material's flexibility. Young's modulus of steel is 0.21 TPa, which is 30,000 times larger than rubber. Therefore, steel is less flexible than rubber. The Young's modulus of CNTs is between 1.28 to 1.8 TPa, which is almost 10 times greater than that of steel, indicating that CNTs are very rigid and tough to bend (Poole and Owens 2003).

Once CNTs are twisted, they are buoyant and bend like straws, but they do not break. They can regain their toughness without any fracture and damage. Due to the presence of defects in the lattice of many other materials in terms of dislocation and grain boundaries, they suffer from bending and fracture, but CNTs have very few defects in their walls, so that they can sustain a tremendous amount of force without any loss. One more reason for the lack of damage when load is applied is that the hexagonal carbons can change their configuration by bending without breaking C=C bonds. Because CNTs hold sp^2 hybridization, after bending they reform sp^2 hybridization again. The amount of reformed sp^2 hybridization between C=C bonds depends on the degree of bending.

With regard to the mechanical properties of CNTs, the strength cannot be defined as stiffness. Young's modulus describes the stiffness and flexibility of a given material. On the other hand, tensile strength defines the quantity of stress required to pull apart a given material. The tensile strength of CNTs is almost 45 billion pascals, while that of high strength steel alloy is less than 2 billion pascals. Consequently, CNTs are 20 times stronger than steel alloys. The SWCNTs are always better than nested CNTs, whereas nested CNTs also have greater mechanical strength, such as MWCNTs, with a diameter of 200 nm exhibiting Young's modulus of 0.6 TPa and a tensile strength of 0.007 TPa (Poole and Owens 2003).

4.4.3 Thermal Properties

CNTs have very high thermal conductivity, which is significantly higher than the diamond. They are excellent conductors of heat. The heat conductivity and thermal expansion can be explained based on the topology of a single layer of graphene in graphite and CNTs. To fold a graphene sheet around a particular axis for developing CNTs, the radial expansion is generated in whole CNTs by covalent bonds between carbon atoms, due to which the van der Waals interactions between nested tubes results in radial thermal expansion. It is assumed that, in defect-free CNTs, the thermal expansion coefficient will be isotropic. This isotropic property of the thermal expansion coefficient may be more beneficial in carbon–carbon composites. With the help of this property, carbon fibres can be contracted or expanded considerably more radially than longitudinally on cooling and heating (Rellick 1990). In the same manner, a carbon matrix can exhibit thermal expansion in CNTs and in graphite, resulting in detrimental stress-induced fracture damage. This problem can be resolved by substituting CNTs with carbon fibres in composites. Defect-free CNTs exhibit a shallow thermal expansion coefficient value, which can generate undesirable stress in different polymers and epoxy resins during the formation of composites (Ruoff and Lorents 1995).

4.4.4 Vibrational Properties

Carbon atoms vibrate between the ellipse and the sphere. The vibrational properties are generated in the molecule due to the vibration of atoms back and forth or up and down. Every molecule contains a set of vibrations known as the normal mode of vibration, which is determined by the molecule's symmetry. CNTs also have normal mode of vibration, as shown in Figure 4.5. One mode, namely A1g (Figure 4.5a), includes in-and-out oscillation of the diameter of CNTs. One other mode, labelled E2g (Figure 4.5b), comprises the pressing-in of CNT on one side and enlarging in the perpendicular direction. The frequencies of both modes depend on CNT diameter and Raman active mode (Poole and Owens 2003).

4.4.5 Elastic Properties

It has been considered that the C=C bond of any substance is the strongest bond in nature. Hence, CNTs possess strength in the direction of the basis vector due to

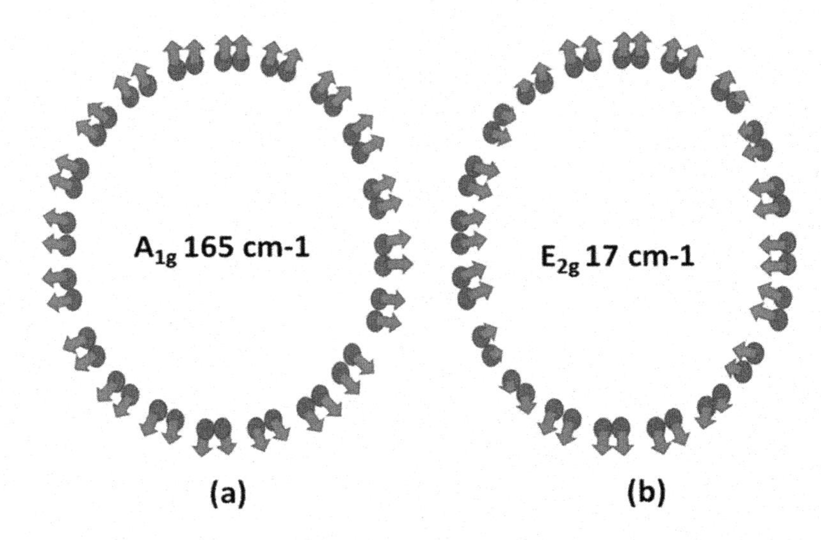

FIGURE 4.5 Schematic of two modes of vibration of CNTs.

the presence of the C=C bond. Additionally, SWCNTs are more elastic and flexible in a direction normal to the surface of the CNTs, due to three fundamental forces between C-C atoms, graphite, and CNTs exhibiting specific elastic properties. These fundamental forces comprise strong σ-bond and л-bond forces between intralayer C=C bonds and weak interlayer interactions. Though these three forces can be distinguished from each other on the basis of their degree of magnitude, even they play a crucial role in determining the complete elastic properties of CNTs. To design a SWCNT, a graphene sheet is folded in a particular direction, resulting in an increase in the strain energy accompanying the curvature of the SWCNT. Therefore, the total energy of SWCNT is increased due to the strain energy. This strain energy is also enhanced by reducing the diameter of the SWCNT. On decreasing the size of the diameter of SWCNT, its stability decreases significantly.

In the graphene layer, sp^2 covalent bonds are responsible for developing the σ-bond of the honeycomb matrix. This honeycomb matrix is considered to be a thin, elastic film over which standard elastic models can be applied (Yakobson et al. 1996). It is assumed that the surface of SWCNTs is relatively soft for applying a force perpendicular to the surface. Considering a tangential force subjected to the σ-bond can illuminate this feature of SWCNTs accurately. Subsequently, inside every SWCNT is a hollow vacuum core. Therefore, there is no interaction between interlayers of SWCNT. A marked bending is obtained by applying a force perpendicular to the axis of SWCNTs without destroying the σ-bond matrix. Although this perpendicular bending or deformation is not always permanent, it results in easy flattening of the cross-section of SWCNTs (Yakobson et al. 1996, Ruoff and Lorents 1995).

On the other hand, MWCNTs exhibit several fascinating properties that are not investigated or defined with the help of the same physics as SWCNTs. The comparatively weak interaction between successive interlayers governs their corresponding stacking structure in the MWCNTs, due to which the position of atomic sites can be defined more precisely on the external layer than on the inner layer. The lattice structure of the external and inner layers are usually inadequate with each other, which defines the turbostratic structure of MWCNTs. The existing shear stress between CNT shells can be affected by this turbostratic structure. On the other hand, a very effective correlation often occurs between stacking layers of MWCNTs. This faceting feature in some MWCNTs has also been demonstrated experimentally (Zhang et al. 1993, Liu and Cowley 1994).

4.5 MODIFICATION

It is well known that CNTs are chemically inactive. Thus, under some circumstances, chemical reactions have to be carried out for significant chemical modification, which may cause structural damage to CNTs, which may destroy its basic properties (Karousis et al. 2010, Tasis et al. 2006). In this situation, physical modification of CNTs is always favourable in preserving electronic properties because of the non-covalent approach for various applications. Using physical modification techniques, like ultra-sonication for even distribution of CNTs in any matrix, CNTs may fracture.

This technique also has some drawbacks, such as inadequate modification and incomplete material consumption during the process. Therefore, the development of a non-damaging and efficient chemical modification technique is required. Numerous modification techniques of CNTs, with their drawbacks and benefits, have been tabulated in Table 4.1.

In 2004, Baek et al. (2004) described pioneering work to create permanent substituted sites of sp^2 hybridized C–H, using direct Friedel-Crafts acylation reactions on the surface of CNTs. Subsequently, their team and other researchers have widely investigated (Choi et al. 2007, Han et al. 2008, 2009, Jeon et al. 2008, 2010, Lee et al. 2008) this technique for various applications because of its many advantages like adequate modification stages, non-destructive reaction nature, utilization of miscellaneous materials, and suitability for mass production.

Furthermore, chemical oxidation in the presence of strong acids can modify the surface of CNTs (Karousis et al. 2010, Tasis et al. 2006). Hence, as a result of chemical oxidation, serious structural damage can be obtained in CNTs. As a consequence, changes can also be observed in the valuable intrinsic properties of CNTs. To resolve these problems, a new functionalization route has been developed, in which homogeneous surface functional groups and several foreign matrixes have been introduced at high density. This route can enhance the compatibility of CNTs with functional groups and reduce the structural damage to CNTs during reactions to improve their properties in highly demanding applications. Baek and colleagues, including many researchers (Choi et al. 2007, Hans et al. 2008, 2009, Jeon et al. 2010, Lee et al. 2008), reported this effective route to covalently functionalize CNTs *via* a modest reaction called direct Friedel-Crafts acylation. Carboxylic acid chloride (COCl), simple benzoic acid (–COOH) and benzamide ($-CONH_2$) groups are directly utilized in this Friedel-Crafts acylation strategy of functionalization. The

TABLE 4.1
Numerous Modification Techniques of CNTs With Their Drawbacks and Benefits (Ma et al. 2010)

Technique	Approach	Principle	Possible Fracture	Easy to Use	Interaction With Polymer Matrix	Agglomeration of CNTs in Matrix
Physical	Endohedral method	Capillary effect	No	No	Weak	Yes
	Surfactant adsorption	Physical adsorption	No	Yes	Weak	No
Chemical	Polymer wrapping	Van der Waals force, Stacking	No	Yes	Varying	No
	Defect	Defect alteration	Yes	Yes	Strong	Yes
	Side wall	sp^2 to sp^3 hybridization	Yes	No	Strong	Yes

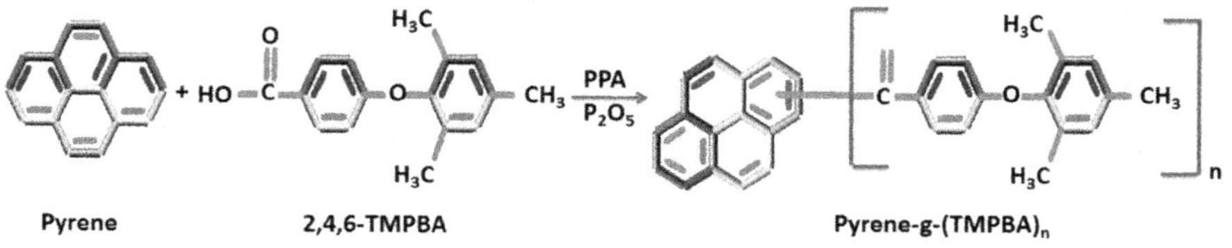

FIGURE 4.6 A summary of reaction of direct Friedel-Crafts acylation reaction using pyrene as a model compound in poly(phosphoric acid)/phosphorus pentoxide medium.

comprehensive reaction mechanism of this Friedel-Crafts acylation strategy, using pyrene as a classical compound, is illustrated in Figure 4.6 (Jen et al. 2010). This reaction usually proceeds in a weak medium of polyphosphoric acid (PPA)/phosphorus pentoxide (P_2O_5) between CNTs and benzoic acid derivatives.

There are two essential roles of PPA for use in this reaction. The first role of PPA says that it is a viscous polymeric acid (pKa ≈ 2.1), which could be capable of protonating the surface of CNTs without causing structural damage in the separation process. In this way, the exceptional electrical, mechanical and thermal properties of CNTs can be conserved. Moreover, the dual role of this viscous nature of PPA is to prevent re-accumulation of CNTs after distribution using mechanical stirring with strong shear forces. An additional component (P_2O_5) of the reaction works as a dehydrating agent to stimulate the Friedel-Crafts reaction effectively. Under these reaction circumstances, imperfect inherent sp^2 hybridized C–H groups are available on the outer shell of CNTs, which are very active locations for conducting electrophilic replacement reactions by freshly produced carbonium ions (C=O+) from benzoic acid and benzylamide by-products in PPA/P_2O_5 (Jeon et al. 2008). Subsequently, an efficient and identical introduction of numerous functional groups without structural deformation of the CNTs was achieved by efficiently establishing direct Friedel-Crafts acylation in PPA/P_2O_5.

Moreover, modest and accessible features of this approach could be considered to be supplementary advantages. The weight ratio of PPA/P_2O_5 has been optimized as 4:1 in the reaction matrix. The reaction is carried out by mechanical stirring using high-torque under dry nitrogen purge for 48–72 h at 130°C. In later runs of this reaction, the solid is transferred to an extraction thimble and extracted with water for 3 days and methanol for 3 days, and finally freeze-dried for 48 h to achieve absolute products (Chang et al. 2013).

4.6 REFERENCES

Ajayan, P. M. and Lijima, S. 1992. Smallest carbon nanotube. *Nature* 358:23–23.

Ajayan, P. M., Ebbesen, T. W., Ichihashi, T., Iijima, S., Tanigaki, K. and Hiura, H. 1993. Opening carbon nanotubes with oxygen and implications for filling. *Nature* 362:522–525.

Baek, J. B., Lyons, C. B. and Tan, L. S. 2004. Covalent modification of vapour-grown carbon nanofibers via direct Friedel–Crafts acylation in polyphosphoric acid. *Journal of Materials Chemistry* 14:2052–2056.

Bethune, D. S., Kiang, C. H., De Vries, M. S., Gorman, G., Savoy, R., Vazquez, J. and Beyers, R. 1993. Cobalt-catalysed growth of carbon nanotubes with single-atomic-layer walls. *Nature* 363:605–607.

Chang, D. W., Jeon, I-Y., Choi, H-J and Baek, J-B. 2013. *Mild and Nanodestructive Chemical Modification of Carbon Nanotubes (CNTs): Direct Friedel-Crafts Acylation Reaction in Physical and Chemical Properties of Carbon Nanotubes.* ed. S. Suzuki, 295–318. Croatia: Intech.

Choi, J. Y., Oh, S. J., Lee, H. J., Wang, D. H., Tan, L. S. and Baek, J. B. 2007. In-situ grafting of hyperbranched poly (ether ketone) s onto multiwalled carbon nanotubes via the A3+ B2 approach. *Macromolecules* 40:4474–4480.

Dravid, V. P., Lin, X., Wang, Y., Wang, X. K., Yee, A., Ketterson, J. B. and Chang, R. P. 1993. Buckytubes and derivatives: their growth and implications for buckyball formation. *Science* 259:1601–1604.

Ebbesen, T. W. 1994. Carbon nanotubes. *Annual Review of Materials Science.* 24:235–264.

Ebbesen, T. W. 1996. *Carbon Nanotubes: Preparation and Properties.* Princeton, NJ: CRC Press.

Ebbesen, T. W. and Ajayan, P. M. 1992. Large-scale synthesis of carbon nanotubes. *Nature* 358:220–222.

Ebbesen, T. W., Hiura, H., Fujita, J., Ochiai, Y., Matsui, S. and Tanigaki, K. 1993. Patterns in the bulk growth of carbon nanotubes. *Chemical Physics Letters* 209:83–90.

Endo, M., Takeuchi, K., Igarashi, S., Kobori, K., Shiraishi, M. and Kroto, H. W. 1993. The production and structure of pyrolytic carbon nanotubes (PCNTs). *Journal of Physics and Chemistry of Solids* 54:1841–1848.

Endo, M., Takeuchi, K., Kobori, K., Takahashi, K., Kroto, H. W. and Sarkar, A. 1995. Pyrolytic carbon nanotubes from vapor-grown carbon fibers. *Carbon* 33:873–881.

Han, S. W., Oh, S. J., Tan, L. S. and Baek, J. B. 2008. One-pot purification and functionalization of single-walled carbon nanotubes in less-corrosive poly (phosphoric acid). *Carbon* 46:1841–1849.

Han, S. W., Oh, S. J., Tan, L. S. and Baek, J. B. 2009. Grafting of 4-(2, 4, 6-trimethylphenoxy) benzoyl onto single-walled carbon nanotubes in poly (phosphoric acid) via amide function. *Nanoscale Research Letters* 4:766.

Iijima, S. 1991. Helical microtubules of graphitic carbon. *Nature* 354:56–58.

Iijima, S. 1993. Growth of carbon nanotubes. *Materials Science and Engineering: B* 19:172–180.

Iijima, S. and Ichihashi, T. 1993. Single-shell carbon nanotubes of 1-nm diameter. *Nature* 363:603–605.

Ikazaki, F. et al. 1994. Chemical purification of carbon nanotubes by use of graphite intercalation compounds. *Carbon* 32:1539–1542.

Jeon, I. Y., Lee, H. J., Choi, Y. S., Tan, L. S. and Baek, J. B. 2008. Semimetallic transport in nanocomposites derived from grafting of linear and hyperbranched poly (phenylene sulfide) s onto the surface of functionalized multi-walled carbon nanotubes. *Macromolecules* 41:7423–7432.

Jeon, I. Y., Tan, L. S. and Baek, J. B. 2008. Nanocomposites derived from in situ grafting of linear and hyperbranched poly (ether-ketone) s containing flexible oxyethylene spacers onto the surface of multiwalled carbon nanotubes. *Journal of Polymer Science Part A: Polymer Chemistry* 46:3471–3481.

Jeon, I. Y., Choi, E. K., Bae, S. Y. and Baek, J. B. 2010. Edge-functionalization of pyrene as a miniature graphene via Friedel–Crafts acylation reaction in poly (phosphoric acid). *Nanoscale Research Letters* 5:1686.

Jeon, I. Y., Kang, S. W., Tan, L. S. and Baek, J. B. 2010. Grafting of polyaniline onto the surface of 4-aminobenzoyl-functionalized multiwalled carbon nanotube and its electrochemical properties. *Journal of Polymer Science Part A: Polymer Chemistry* 48:3103–3112.

Journet, C. et al. 1997. Large-scale production of single-walled carbon nanotubes by the electric-arc technique. *Nature* 388:756–758.

Karousis, N., Tagmatarchis, N. and Tasis, D. 2010. Current progress on the chemical modification of carbon nanotubes. *Chemical Reviews* 110:5366–5397.

Kiang, C. H., Goddard III, W. A., Beyers, R. and Bethune, D. S. 1995. Carbon nanotubes with single-layer walls. *Carbon* 33:903–914.

Lee, H. J., Han, S. W., Kwon, Y. D., Tan, L. S. and Baek, J. B. 2008. Functionalization of multi-walled carbon nanotubes with various 4-substituted benzoic acids in mild polyphosphoric acid/phosphorous pentoxide. *Carbon* 46:1850–1859.

Liu, M. and Cowley, J. M. 1994. Structures of carbon nanotubes studied by HRTEM and nanodiffraction. *Ultramicroscopy* 53:333–342.

Ma, P. C., Siddiqui, N. A., Marom, G. and Kim, J. K. 2010. Dispersion and functionalization of carbon nanotubes for polymer-based nanocomposites: a review. *Composites Part A: Applied Science and Manufacturing* 41:1345–1367.

O'Connell, M. J. 2006. *Carbon Nanotubes: Properties and Applications*. Princeton, NJ: CRC Press.

Poole Jr, C. P. and Owens, F. J. 2003. *Introduction to Nanotechnology*. Hoboken, NJ: Wiley.

Rao, A. M. et al. 1997. Diameter-selective Raman scattering from vibrational modes in carbon nanotubes. *Science* 275:187–191.

Rellick, G. 1990. Densification efficiency of carbon-carbon composites. *Carbon* 28:589–594.

Ruoff, R. S. and Lorents, D. C. 1995. Mechanical and thermal properties of carbon nanotubes. *Carbon* 33:925–930.

Ruoff, R. S. and Lorents, D. C. 1995. Mechanical and thermal properties of carbon nanotubes. *Carbon* 33:925–930.

Saito, R., Dresselelaus, G. and Drbselmus, M. S. 1998. *Physical Properties of Carbon Nanotubes*. London: Imperial College Press.

Saito, Y., Yoshikawa, T., Inagaki, M., Tomita, M. and Hayashi, T. 1993. Growth and structure of graphitic tubules and polyhedral particles in arc-discharge. *Chemical Physics Letters* 204:277–282.

Smalley, R. E. 1992. Self-assembly of the fullerenes. *Accounts of Chemical Research* 25:98–105.

Tasis, D., Tagmatarchis, N., Bianco, A. and Prato, M. 2006. Chemistry of carbon nanotubes. *Chemical Reviews* 106:1105–1136.

Thess, A. et al. 1996. Crystalline ropes of metallic carbon nanotubes. *Science* 273:483–487.

Tsang, S. C., Harris, P. J. F. and Green, M. L. H. 1993. Thinning and opening of carbon nanotubes by oxidation using carbon dioxide. *Nature* 362:520–522.

Yakobson, B. I., Brabec, C. J. and Bernholc, J. 1996. Nanomechanics of carbon tubes: instabilities beyond linear response. *Physical Review Letters* 76:2511.

Yudasaka, M., Komatsu, T., Ichihashi, T. and Iijima, S. 1997. Single-wall carbon nanotube formation by laser ablation using double-targets of carbon and metal. *Chemical Physics Letters* 278:102–106.

Zhang, X. F., Zhang, X. B., Van Tendeloo, G., Amelinckx, S., De Beeck, M. O. and Van Landuyt, J. 1993. Carbon nano-tubes; their formation process and observation by electron microscopy. *Journal of Crystal Growth* 130:368–382.

5 Graphene Nanoribbons, Fabrication, Properties, and Biomedical Applications

Asha P. Johnson, H. V. Gangadharappa, and K. Pramod

CONTENTS

DOI: 10.1201/9781003110781-5

5.1 INTRODUCTION

Graphene, a single layer of graphite composed of sp^2 hybridized carbon, has attracted extensive research interest in diverse fields since its first discovery in 2004 due to its combination of unique physicochemical, electronic, mechanical, optical, and thermal properties (Cataldo et al. 2010; Yang et al. 2018). It is the strongest, thinnest, and lightest material known (Zhang et al. 2016). Graphene is the basic building block of other carbon nanomaterials of different dimensionalities, such as graphene quantum dots (GQDs), carbon nanotubes (CNTs), graphene nanoribbons (GNRs), three-dimensional graphite, etc. (Liao et al. 2018; Nakano et al. 2018) (Figure 5.1a). Since 2004, much research has been conducted on graphene and its potential applications in fields such as biomedical (Jaleel et al. 2019; Mauri et al. 2021), sensing (Niu et al. 2021; Rabchinskii et al. 2021), engineering (Bei et al. 2019; Naghdi et al. 2020), and energy storage (Folorunso et al. 2020; Pramod Kumar et al. 2020).

Graphene nanoribbons (GNRs) are strips of graphene, just a few nanometres wide, and represent a new graphene derivative class with potential applications in various fields (Kumar et al. 2019b). GNRs enjoy edges and a high surface-to-volume ratio (Jaiswal et al. 2015). One of the most memorable characters of GNRs is the heightened sensitivity to their edge characteristics and dimensions (Hou and Yee 2007a). GNR shares the fascinating features of graphene with exceptional tunable optoelectronic properties. Conventional implementation of graphene in electronics is restricted by the difficulty of the bandgap opening (Castellanos-Gomez and Jan van Wees 2013). GNRs can overcome this limitation by opening up bandgap in the form of lengthened strips of graphene (Lu and Guo 2010). GNRs exhibit invaluable electronic properties, from semiconducting to metallic, based on their edge characteristics and dimensions (Yazyev 2013). GNRs are attractive as a distinctive material applicable to various research fields due to their amazing combination of electronic, mechanical, optical, and thermal properties, ultra-high surface area, and edge characteristics.

Graphene oxide nanoribbons (GONRs) are oxygenated graphene derivatives that exhibit biomedical application potential even greater than that of GNRs due to their various oxygen functionalities which can be exploited by attaching biomolecules. GONRs are amphiphilic graphene oxide derivatives. GONRs are more suitable for drug delivery and gene delivery applications, whereas non-oxidized GNRs are more appropriate for sensing applications (Mousavi et al. 2019). Surface modification of GNRs is considered to be a vital step in health care applications, making them more biocompatible. Covalent and non-covalent types of modification are possible with GONRs. The π electron system allows easy non-covalent conjugation with aromatic parts of molecules for surface modifications, and oxygenated functionalities are suitable for making a strong covalent bond with molecules (Genorio and Znidarsic 2014).

GNRs are emerging as a potential platform for biomedical applications. Biosensing is one of the most studied health care applications of GNRs. High thermal and electrical conductivity, high surface area, high mechanical strength, optical properties, and provision for covalent and non-covalent modifications make them an attractive platform for signal transduction in biosensing. The π–π interactions between GNRs and drugs of an aromatic nature allow easy encapsulation of drugs for an efficient delivery system (Chowdhury et al. 2015b). Anticancer therapy, gene therapy, antimicrobial activity, photothermal therapy, bone regeneration, locomotor function recovery, etc., are among the important applications of GNRs in health care.

5.2 TYPES OF GNRs

Different edge configurations are formed during the production of GNRs. GNRs are classified mainly into two edge configurations, zig-zag GNRs (ZGNRs) and armchair GNRs (AGNRs) (Kusuma et al. 2018; Wakabayashi 2012). Zig-zag edges are triangular edges of hexagons, whereas armchair edges are made up of hexagonal sides (Celis et al. 2016). GNRs with alternate zig-zag and armchair edges are known as chiral GNRs (CGNRs). ZGNRs exhibit metallic behaviour, and their Fermi energy is confined to the flat band. On the other hand, AGNRs have shown metallic as well as semiconducting behaviour. The behavioural change of AGNRs is based purely on their width (Pefkianakis et al. 2015; Xie et al. 2011). Due to the intrinsic magnetic ground state, ZGNRs are preferred over AGNRs. The electronic characteristics of GNRs are based on the edge termination pattern. The highly reactive nature of ZGNRs causes the reconstruction of bare ZGNRs (Jaiswal et al. 2015). Concerning the cutting direction, both ZGNR and AGNR have a 30 ° orientation difference, which will change the π electronic structure and make a localized edge state for ZGNRs. But edge state does not exist in AGNRs (Janani and Thiruvadigal 2018a). A schematic illustration of the design of the different types of GNR is depicted in Figure 5.1b.

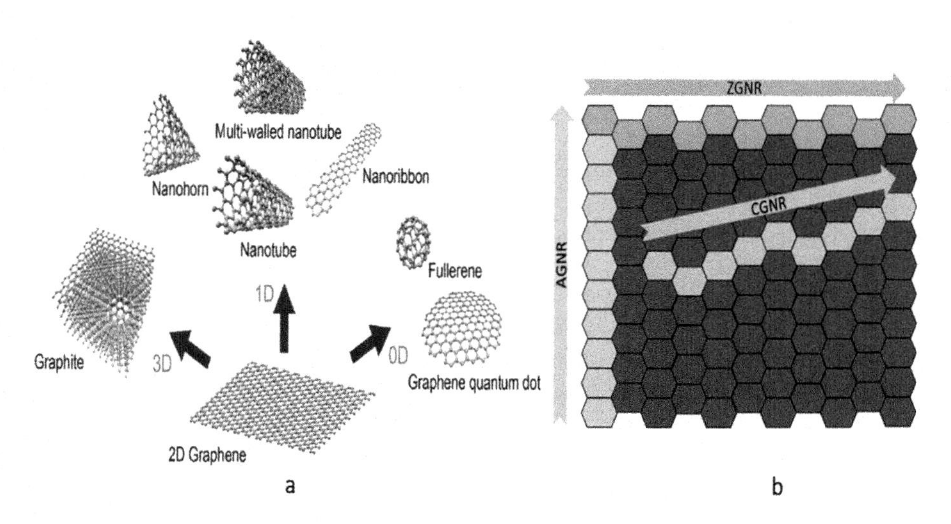

FIGURE 5.1 a) Graphene derivatives of different dimensionalities (Nakano et al. 2018); b) schematic illustration of types of GNRs.

5.3 PROPERTIES OF GRAPHENE NANORIBBONS

GNRs are lengthened strips of graphene composed of sp^2 hybridized carbons. GNRs have a high length-to-width ratio and their size is expressed by their length and width measurements (Bu et al. 2009). AGNR width is determined by the number of hexagonal rings (generally known as dimer lines (Na)), whereas, for ZGNRs, the number of the zig-zag chains (Nz) determines the width (Sharda and Agarwal 2014). Structurally, GNRs are π-conjugated systems that resemble a honeycomb lattice structure. The ultra-high surface area offers excellent characteristics in the biomedical field (Gundra and Shukla 2011). GNR surfaces are inert like graphene, and they can form physical interactions with other molecules (π-π conjugations) (Kan et al. 2008). The characteristics of GNRs are mainly determined by their edge pattern (Zhu et al. 2009). The oxidized form of GNRs is known as graphene oxide nanoribbons (GONRS), with different oxygen functionalities like carboxyl, carbonyl, epoxide, and hydroxide on the surface GONRs. Hence, GONRs show greater potential than GNRs in the biomedical field. Abundant oxygen functionalities increase the functionalization potential as well as the biomedical applications. GONRs are amphiphilic graphene derivatives that can form aqueous dispersions and organic dispersions due to the presence of oxygen functionalities and π conjugations, respectively (Rajaji et al. 2018). The physical properties of GNRs depend significantly on the size and number of layers, which, in turn, depend on the GNR synthesis method (Dimiev et al. 2018).

Exceptionally high charge carrier mobility is the most attractive graphene character, making it a potential material for electronic applications. But the absence of bandgap restricts the use of graphene in logical devices. Making graphene in lengthened strips is the best approach to tuning the electronic properties by opening up the bandgap. GNRs are a quasi-one-dimensional graphene layer that allows a controllable electronic character by changing its edge shape and width (Tomita and Nakamura 2013). GNRs possess peculiar electronic structure and transport characteristics (Hou and Yee 2007b), while ZGNRs show a localized edge state with energies close to the Fermi level.

Conversely, AGNRs do not keep such a localized edge state (Wakabayashi 2012). The formation of bandgap in GNRs is mainly based on width, edge pattern, electron confinement, strain, and doping (Corso et al. 2018). The bandgap of ZGNRs increases with increasing GNR width whereas it is inversely proportional in AGNRs (Berahman et al. 2015). When the number of dimer lines is 3 Na or 3 Na+1, the AGNRs are semiconducting, whereas, if it is 3 Na+2, it is metallic. Irrespective of the Nz, ZGNRs are always metallic (Sharda and Agarwal 2014). When GNRs are in a multi-layered form, ZGNRs have more edge states and AGNRs have lower band gaps.

GNRs exhibit some excellent thermal characteristics owing to their edge and low dimensionality. GNRs are a potential next-generation heat dissipation material due to their high thermal conductivity (Tomita and Nakamura 2013). The phonon thermal conductance of GNRs is higher than that of 2-dimensional graphene, and the conductance of GNRs is sensitive to its number of atoms (width). In ZGNRs, the thermal conductivity increases and then decreases with an increase in the number of atoms. But a consistent growth occurs in the case of AGNRs (Guo et al. 2009). The ZGNRs exhibit greater thermal conductance than single-walled carbon nanotubes (SWCNTs) (Tomita and Nakamura 2013). Over the entire temperature range, thermal conductivity enhances with the length of the GNRs. At room temperature, the heat transport anisotropy of GNRs suggests that AGNRs possess lower thermal conductivity than ZGNRs, though, above 100 nm width, the anisotropy disappears (Johnson et al. 2020). GNRs can show outstanding ballistic transport in the presence of substrate-induced disorder, but only ZGNRs can retain this ballistic nature with edge disorder (Watanabe et al. 2009).

AGNRs can absorb photons from terahertz to infrared and visible radiation. Thus, AGNRs are good candidates for photodetectors. The edge pattern is the primary factor in controlling the optical behaviour of GNRs. Modification of the edge pattern allows the fine-tuning of optical characteristics (Balarastaghi et al. 2016). Chiral GNRs show improved optical performances and accurate selectivity within the optical spectrum. By adjusting the edge parameters, the optical absorbances can be controlled (Berahman et al. 2015). The selection rule for AGNRs and ZGNRs are different; width and edge properties are two significant factors that play an important role in the selection rule (Hsu and Reichl 2007).

The mechanical characteristics of GNRs are determined by their size, edge, and chirality. Based on their width and chirality, the strength will range from 90 to 180 GPa (Orlov and Ovid'ko 2015). Density functional theory calculations showed that size-induced effects are different for AGNRs and ZGNRs. Size-induced effects do not significantly influence Young's modulus or strength of AGNRs, but are significantly associated with Poisson's ratio. For ZGNRs, Young's modulus and strength will increase with a decrease in the width of GNRs. The density functional theory (DFT) results suggest that edge defects reduce the maximum strength of GNRs, and chirality becomes significant only at high strains (Tabarraei et al. 2015). ZGNRs have strong magnetic properties because they possess a localized edge state with energy near the Fermi level (Wakabayashi 2012). The edge-induced magnetic property of GNRs is precarious under normal environmental conditions.

5.4 SYNTHESIS OF GRAPHENE NANORIBBONS

The synthesis of atomically precise, high-quality, narrow GNRs is critical for several GNR applications. Generally, two alternative methods are used for the fabrication of GNRs, namely the top-down and bottom-up methods. The top-down method involves fabricating GNRs from larger blocks of sp^2 carbon nanostructures, such as carbon nanotubes or graphites. The bottom-up method involves utilizing small molecules as building blocks to make lengthened strips of graphene.

5.4.1 Top-down Methods

The top-down method includes nanotube unzipping, lithographic patterning, sonochemical cutting, and metal-catalyzed cutting and etching of graphene. Top-down methods generally yield nanoribbons of width greater than 10 nm and have limited control over the edge pattern (Casiraghi and Prezzi 2017). Oxidative unzipping is the most widely used technique for synthesis, but few studies have been reported for GNRs of width less than 10 nm.

5.4.1.1 Unzipping of Nanotubes

The unzipping of nanotubes is the most commonly used method for the synthesis of GNRs. Carbon nanotubes are a rolled-up form of graphene sheets. GNRs exhibit more structural flexibility than CNTs (Wang et al. 2019), so the unzipping or unravelling of single-walled carbon nanotubes (SWCNTs) or multi-walled carbon nanotubes (MWCNTs) can produce GNRs (Terrones et al. 2010). Several methods have been used for breaking C–C bonds to achieve unzipping or unravelling of carbon nanotubes, such as oxidation cutting, electrical unzipping, intercalation exfoliation, laser-induced unzipping, plasma etching, electrochemical unzipping, high-impact collision, or potassium vapour-mediated unzipping.

CNT unzipping involves the reaction of MWCNTs in acidic oxidative media. The oxidative unzipping method for MWCNTs is similar to the Hummers method used to prepare graphene oxide from graphite (Dimiev et al. 2018). Strong oxidants like potassium permanganate and sulphuric acid, are used in most oxidative unzipping reactions. These can weaken the C–C bond. Research shows that oxidation results in wrinkled structures, which can be cut into small pieces by atomic force microscopy (Fujii and Enoki 2010). This method produces water-soluble GONRs. The oxidative unzipping happens along a line that leads to the formation of straight-edged GONRs. Depending upon the position of the initial attack and the chiral angle of the CNTs, the opening could happen in a spiralling manner or a linear longitudinal cut (Kosynkin et al. 2009). Production of GNRs with a comparatively high yield is the significant primary advantage of this technique, but it is not easy to control the physical and chemical parameters of the synthesized GNRs. Oxidative unzipping is the most commonly used method due to its high throughput and low cost.

Transitional metals can act as a catalyst for the unzipping of nanotubes. The metal particle catalyst breaks the C–C and H–H bonds and unzips the nanotubes. This method is free of aggressive chemicals, and, thus, it produces GNRs with smooth edges (Ma et al. 2012). This method can be carried out with most transition metals, but Fe is the best catalyst since it can unzip the nanotubes at room temperature (Chen et al. 2016). Transitional metal-induced cutting of graphene sheets can also be used to produce GNRs (Solís-Fernández et al. 2013).

The polymer-protected plasma etching method involves the unzipping of carbon nanotubes on exposure to Ar plasma. The polymer is used to protect the carbon nanotubes, and the unprotected part is exposed to unzipping. The polymer film will act as a protective mask and allow the exposed part to unzip (Jiao et al. 2010a).

Intercalation-assisted unzipping is based on the intercalation ability of alkali metals. Lithium and potassium are commonly used intercalants for unzipping. The intercalated metals will expand the carbon nanotubes and induce unbearable stress on the concentric walls, causing the longitudinal unzipping of nanotubes. The resulting GNRs are multi-layered and highly conductive (Dimiev et al. 2018). Thinner and shorter GNRs can be synthesized by laser irradiation unzipping of CNTs. Edge roughness is the principal disadvantage of GNRs formed by laser irradiation (Kumar et al. 2011; Marković et al. 2016). Electrochemical unzipping is another method of making GNRs in the presence of an interfacial electric field. This method offers advantages like manageable layer thickness and orientation (Shinde et al. 2011). Unzipping of CNTs by high-energy mechanical impact on solid objects unzips the tubes in a one-step, chemical-free method for production of GNRs. Higher-velocity impact of CNTs against a solid target can create defects and rapid evaporation of atoms, leading to the longitudinal unzipping of nanotubes (Ozden et al. 2014).

5.4.1.2 Lithographic Patterning

Large arrays of aligned GNRs can be fabricated with lithographic patterning, but it is difficult to control the width and edge smoothness (Jiao et al. 2010b). Lithographic patterning is not suitable for large-scale production of GNRs due to its tedious nature. GNRs of a significantly smaller pattern (10 nm width) can be fabricated by ice-assisted electron beam lithography. This method involves the electron beam-induced cleavage of graphene flakes supported on an ice layer(Gardener and Golovchenko 2012).

5.4.1.3 Sonochemical Cutting

Sonochemical cutting of GNRs produces narrow and lengthy GNRs. This method mainly consists of two processes: chemical exfoliation of graphite to form sheets of graphene followed by sonochemical breakage of the sheets to nanoribbons. The formation of single-layered GNRs and comparatively large yield are the significant advantages associated with this method, but the orientation of the edge is difficult to control with this method (Wu et al. 2010).

5.4.2 Bottom-up Fabrication of Graphene Nanoribbons

The bottom-up synthesis of GNR through surface-assisted polymerization and solution-based synthesis is considered the best method for designing atomically precise GNRs. The rate of success of this method is entirely dependent on the rational design of proper reactants. GNRs with well-defined edge structures, width, and heterojunctions can be fabricated in successful bottom-up methods (Afonso and Palenzuela 2019). Selective growth of building blocks can be controlled to achieve the desired width and edge structure.

5.4.2.1 Solution-based Synthetic Approaches

Atomically precise GNR synthesis has been achieved with several solution-based synthetic techniques. This type of GNR synthesis involves the polymerization of designed molecules and the generation of GNRs with the planarization of the resulting polymer. Suzuki cross-coupling as a polymerization reaction is one of the most successful methods for preparing atomically precise GNRs. Yamamoto coupling reactions are an attractive alternative method to Suzuki coupling reactions for making lengthy GNR precursors. The Yamamoto method uses only one halogenated monomer, whereas the Suzuki coupling reaction needs two different monomers. Diels-Alder addition methods can also achieve more efficient polymerization. The planarization of the polymer fabricated *via* different polymerization reactions is the final step for the synthesis of GNRs (Narita et al. 2014; Narita et al. 2019; Pefkianakis et al. 2015; Shekhirev and Sinitskii 2017). The solution-based synthesis produces long GNRs and liquid phase processable nanoribbons.

5.4.2.2 On-the-surface Synthesis

Surface synthesis mainly involves synthesis under ultra-high vacuum (UHV) conditions and chemical vapour deposition (CVD) based on surface synthesis. Thermal-induced polymerization of various monomers on a suitable surface under UHV conditions, followed by high-temperature cyclo-hydrogenation, will produce atomically precise GNRs. On-surface synthesis under UHV conditions creates GNRs and allows the visualization with various techniques like Scanning Tunneling Microscope (STM) and Atomic Force Microscopy (AFM). UHV conditions are essential for the visualization and to suppress the various side reactions. It involves an expensive equipment that generally includes a very small reaction chamber, limiting the scalability of synthesis. Thus, this method is comparatively costly and has limited scalability. These GNR syntheses under UHV produce monolayer GNRs in a limited area (1 cm^2) (Narita et al. 2019; Chen et al. 2016).

Chemical vapour deposition is one of the fundamental ways of carrying out GNR synthesis, allowing control of both width and length. This method involves the sublimation and deposition of monomers on a metallic surface, followed by thermal annealing-induced surface-assisted polymerization. A metallic template actS as the catalyst for monomer decomposition at high temperature (Saraswat et al. 2019; Celis et al. 2016; Narita et al. 2019). Methane gas is the commonly used precursor; acetylene, ethylene, and propane, etc., are also used in CVD methods. Some liquid and solid precursors are also used (Liu and Speranza 2019). The flat deposition of monomers on a surface avoids aggregation problems in solution-based synthesis. The CVD method is the best method to synthesize large-area GNRs for device integration. The GNRs produced from CVD have less bandgap and absorption over visible to IR regions. Hence, it is a good candidate for optoelectronic applications (Narita et al. 2019). The growth of precursors inside the nanotubes is another method of GNR synthesis. In a study, the conversion of coronene molecules inside boron nitride nanotubes (BNNTs) produced GNRs. The width of the GNRs formed was the same as the width of the nanotubes. Formed GNRs can be removed by oxidation (Barzegar et al. 2016). Various top-down and bottom-up approaches are depicted in Figure 5.2.

5.5 MECHANISM

Fabrication of nanoribbons with desired physical and chemical properties requires a thorough knowledge about the entire GNR production process. Oxidative unzipping involves the treatment of carbon nanotubes in an acidic, oxidative media. Manganate ester formation is the initial and rate-limiting step of oxidative unzipping, followed by dione formation. The buttressing effect on ketones further distorts β,γ alkenes

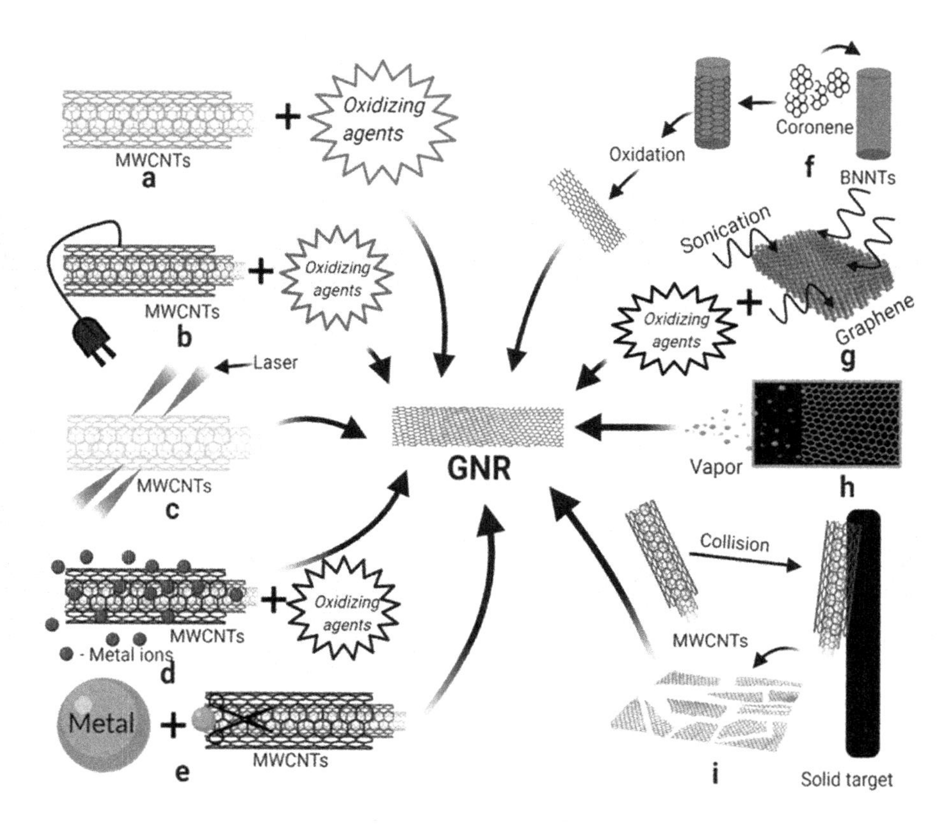

FIGURE 5.2 Top-down and bottom-up approaches of GNR synthesis. a) The longitudinal unzipping of MWCNTs by oxidation cutting; b) electrochemical cutting; c) laser irradiation method; d) intercalation exfoliation; e) metal-catalyzed cutting; f) formation of GNRs inside boron nitride nanotubes; g) sonochemical method; h) chemical vapour deposition (CVD) method; and i) formation of GNRs by high-impact collision of MWCNTs. Reprinted with permission from Johnson, Gangadharappa, and Pramod (2020) © 2020 Elsevier.

and makes them highly prone to permanganate attack. As oxidation continues, strain induced by the buttressing of ketones is reduced due to enhanced space for carbonyl projection, but bond angle strain, caused by the enlarging hole (or tear if originating from the end of nanotube) would make β,γ alkenes more reactive. Once the CNT opening has been started, further opening is increased relative to the unopened tube. The bond angle strain is relieved entirely after opening and the ketones are converted to their oxygen protonated forms (Figure 5.3a) (Kosynkin et al. 2009; Higginbotham et al. 2010).

Dimiev et al. (2018) suggested that the oxidative unzipping mechanism is not achieved by oxidative cleavage of chemical bonds but is driven by intercalation. The mechanism involves three steps: 1) intercalation of graphite by sulphuric acid and formation of the sulphuric acid-graphite intercalation compound; 2) conversion of the intercalation compound into GO (i.e., the formation of C–O bonds); and 3)

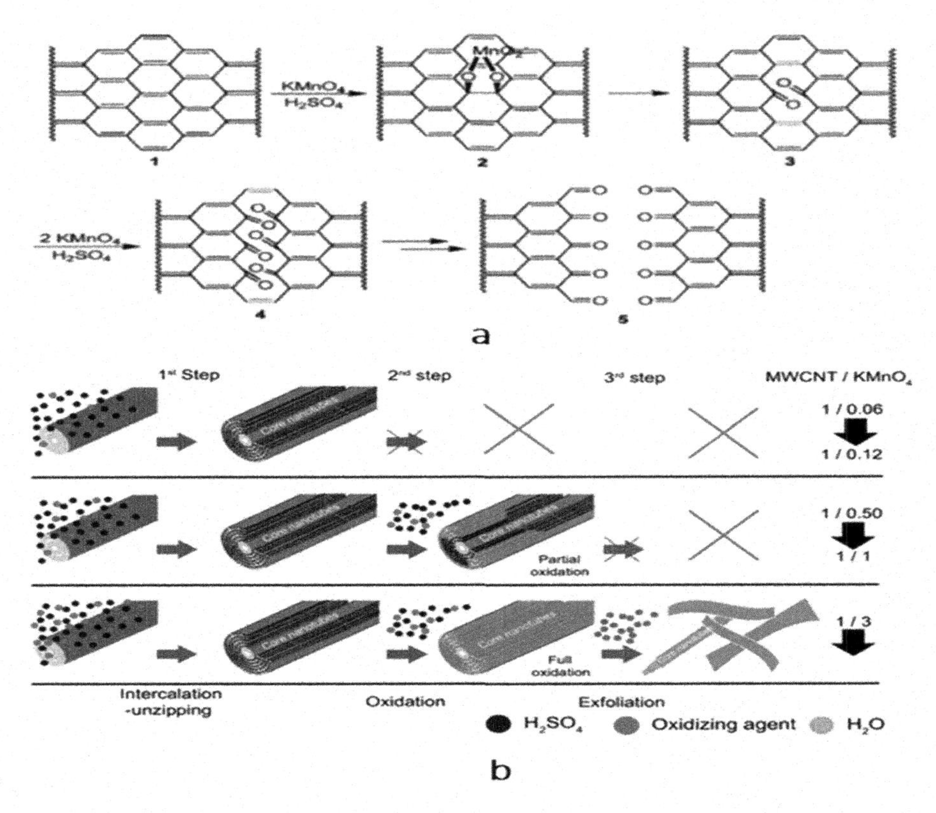

FIGURE 5.3 a) Mechanism of longitudinal unzipping of MWCNTs by oxidative cleavage of C–C bonds. Reprinted with permission from Higginbotham et al. (2010) 2010 © American Chemical Society; b) Intercalation-driven mechanism of unzipping; the grey and orange colours indicate non-oxidative and oxidative conditions, respectively. Reprinted with permission from Dimiev et al. (2018) © 2018 American Chemical Society.

exfoliation of GO sheets to a single layer of GNR on exposure to water (Dimiev et al. 2018). The reaction can extend to different stages, based on their MWCNT/$KMnO_4$ ratio (Figure 5.3b).

The laser irradiation mechanism for unzipping CNTs involves attaining higher temperatures at defective sites than at the non-defective sites of MWCNTs, followed by local heating and initiation of longitudinal unzipping. Little laser energy is required to open up the nanotubes completely. MWCNTs formed by the CVD method become fragmented when the energy is higher than 350 mJ (Kumar et al. 2011). In metal-catalyzed cutting, thermally activated metal nanoparticles at elevated temperatures diffuse to the graphene before meeting the edge. Once they meet the edge, chemically reactive carbon atoms at the edges undergo catalytic hydrogenation by transition metals and lead to methane gas formation. From the reactive edge, metal propagates to the graphene sheet and forms a trench, finally leading to the formation of GNRs (Campos et al. 2009; Solís-Fernández et al. 2013; Chen et al. 2016).

Metal-catalyzed cutting of CNTs involves two stages: reduction of the dissociation barrier of hydrogen molecules and carbon–carbon bond cleavage. The cutting channel along the hydrogen termination of the boundary carbon is critical for the movement of catalyst atoms (Chen et al. 2016; Wang et al. 2011).

5.6 CHARACTERIZATION OF GNRs

GNRs can be characterized by using several spectral and imaging analyses, of which Raman spectroscopy is one of the most important techniques. The spectra provide details about the orientation of layers, the number of layers, disorder, edge type, functionalization, doping, etc. Raman spectroscopy is a high-speed technique, and it can retain the sample without any degradation after spectrum recording. Edge pattern and the relationship with chirality can be identified from the Raman spectra (Shekhirev and Sinitskii 2017; Xie et al. 2011; Casiraghi and Prezzi 2017; Jiao et al. 2010a). The composite analysis and the presence of oxygen can be evaluated by X-ray photoelectron spectroscopy (XPS). The coordination of carbon and oxygen also can be identified by XPS. It is also beneficial in the characterization of heteroatom-doped GNR (Shekhirev and Sinitskii 2017). The number of layers of GNRs that are retained after the unzipping procedures can be identified by fast Fourier transform technology (FFT). Scanning tunnelling microscopy (STM) becomes the most important characterization method in GNR- based research, opening up a direct way to visualize the precise atomic structure of GNRs (Shekhirev and Sinitskii 2017). The identification of morphological features is possible by scanning electron microscopy (SEM) and transmission electron microscopy (TEM) (Xiao et al. 2014; Abdolkarimi-Mahabadi and Manteghian 2015) Atomic force microscopy images allow the identification of height profile and thickness (Zhu et al. 2009; Jiao et al. 2010b) Crystalline nature and thermal behaviour studies are possible with X-ray diffraction (XRD) and thermogravimetric analysis (TGA)(Xiao et al. 2014; Li et al. 2015a). Functional group identification and elemental analysis of GONRs can be conducted by Fourier Transform Infrared (FTIR) spectroscopy and energy-dispersive X-ray spectroscopy (EDX) (Zhu et al. 2009; Govindasamy et al. 2017b).

Spectroscopic techniques are helpful to determine the structural as well as the electron characteristics. GNRs are π-conjugated systems. The degree of π conjugation can be determined by UV-visible spectrophotometry (Shekhirev and Sinitskii 2017). The optical bandgap of GNRs can also be determined from the spectrum. But the spectral characteristics will change according to the measurement conditions. Therefore, to find out the electronic properties, a combination of UV-visible spectroscopy with scanning tunnelling spectroscopy (STS) or photoluminescence (PL) spectroscopy is beneficial. PL spectroscopy is also used to determine the bandgap of GNRs (Shekhirev and Sinitskii 2017).

5.7 FUNCTIONALIZATION OF GNRs

Functionalized graphene derivatives attract tremendous attention in the biomedical field owing to their unique physicochemical properties and biocompatibility.

Functionalization can modify (ideally, improve) the properties of nanomaterials, enabling them for biomedical applications. Several studies have reported that modification of GNRs is an essential step in their health care applications (Yang et al. 2013). GNRs are poorly soluble in most organic solvents due to the π-π interactions. Hence, functionalization of GNRs with solubility-inducing agents is essential. Based on the properties of functional groups, the solubility of GNRs in aqueous or organic solvents may vary (Genorio and Znidarsic 2014). Functionalization of graphene nanomaterials with biocompatible polymers is one of the main modifications in published literature to improve biological performances, stability, and solubility. Covalent and non-covalent changes can be carried out to improve the relevant physicochemical characteristics of GNRs for biomedical applications. The adsorption molecules on graphene can achieve non-covalent functionalization of GNR by π-π stacking or van der Waals forces of attraction (Liu and Speranza 2019). Covalent modification can be achieved by cycloaddition reactions or by the attachment of molecules *via* oxygen functionalities. Covalent attachment is stronger than the non-covalent type of interaction, but non-covalent interactions do not lose the carbon network of GNRs (Reina et al. 2017; Yang et al. 2017).

Modification of GNRs with the amphiphilic polymer DSPE-PEG (1,2-Distearoyl -sn-glycero-3-phosphoethanolamine-Poly(ethylene glycol)) has been reported in various studies of biomedical applications. The hydrophobic part, DSPE, attaches to the aromatic part and forms stable organic dispersions, whereas PEG (poly(ethylene glycol) creates a hydrophilic character. Hydrophobic sphingolipid ceramide was loaded onto GONRs modified with DSPE-PEG. DSPE-PEG improves the nanocarrier's drug loading, aqueous dispersibility, and cytotoxicity natures (Suhrland et al. 2019). In a study, reduced GONR-DSPE-PEG showed higher near-infra-red (NIR) absorption, enhanced targeting, and cytotoxicity than bare GONR and rGO-DSPE-PEG (Akhavan et al. 2012). Due to the π-πinteraction of DSPE-PEG with GONRs, the loading efficiency of genetic materials on GONRs was decreased due to the lower availability of π electrons for loading (Chowdhury et al. 2016).

In GNR-based biosensor fabrications, modifications were performed to improve the sensitivity, selectivity, conductivity, or surface area. Covalent or non-covalent attachment of bio-receptors is possible with GNRs. Metal nanoparticle modification is an efficient strategy to enhance the electrocatalytic activity of GNR-based sensors (Vukojević et al. 2018; Zakaria and Leszczynska 2019; Kumar and Goyal 2018). The formation of hybrid nanomaterials with other graphene derivatives (graphene quantum dots, graphene sheets, MWCNTs, SWCNTs, fullerenes) is popular for its property of surface area enhancement (Lavanya et al. 2017; Li et al. 2017; Fei et al. 2015; Zhao et al. 2019)

Amino acid functionalization and single atom doping can change the physicochemical characteristics and improve the solubility of GNRs. Doping of GNRs with L(+)-leucine and IB elements (copper, silver, gold) showed structural changes, increased ionic energy, and increased aqueous solubility (Janani and Thiruvadigal 2017). L(+)-phenylalanine and boron-doped GNRs also showed structural and electronic character changes and improved aqueous solubility (Janani and John

Thiruvadigal 2018a). Similar results were reported with L(+)-serine, L(+)-valine, and oxygen-functionalized GNRs (Kumar et al. 2019a) (Table 5.1).

5.8 BIOMEDICAL APPLICATIONS OF GNRs

5.8.1 Biosensing

One of the most studied health care applications of GNRs is biosensing. A biosensor consists of two critical parts: a biological recognition part as a receptor and a transducer. The analyte molecules interact with the receptor, and the transducer converts the information regarding the interaction into a readable signal. There are several studies regarding the use of graphene derivatives as physical transducers for signal transduction. The unique properties of GNRs, such as electrical conductivity, thermal conductivity, optical property, high mechanical strength, ultra-high surface area, etc., make them a promising platform for biosensing. Surface modification is a significant part of the design of a biosensor. Covalent and non-covalent modifications with various biological and chemical molecules enhance the sensing property. Nanopore-based detection is another important method in GNR-based detection. Different metallic nanoparticles enhance the conductivity, electrocatalytic activity, and surface area of the GNR-based sensors (Pham et al. 2019). Hybrid sensors with other graphene derivatives can improve the surface area of GNR-based sensors (Lavanya et al. 2017; Zakaria and Leszczynska 2019). Doping and modification with polymers are also used to modify the sensing characteristics of GNR-based biosensors (Rastgoo and Fathipour 2019; Pan et al. 2015).

5.8.1.1 GNR-based Biosensors in Biomarker Detection

An enzyme-based reagent-less biosensor was designed for the accurate detection of glucose by attaching glucose oxidase to unzipped GONR. Flavine adenine nucleotide (FAD)-conjugated GONRs were obtained longitudinally on a screen-printed carbon electrode (SPCE), and the separated apoenzyme was connected to holoenzyme-conjugated FAD. The Nafion membrane-stabilized biosensor showed a good response towards detecting glucose (Mehmeti et al. 2017). Similarly, an ultrasensitive GNR sensor was fabricated with graphene nanoribbon/graphene sheet/nickel nanoparticle (GNR/GS/Ni) hybrid material for glucose detection. The hybrid material offers a high surface area, reduced stacking, and high electrochemical activity. The GNR/GS/Ni hybrid biosensor exhibited a comprehensive linear amperometric response, low detection limit, and high sensitivity towards glucose. This sensor is insensitive to interfering substances commonly found in actual samples (Lavanya et al. 2017).

The electrical conductivity of GNRs has been exploited to develop a nanopore-based biosensor for the detection of DNA. A Z-shaped GNR sensor was developed with two ZGNRs and one AGNR with a centralized nanopore. When nucleobases are translocated through the hydrogen-passivated nanopore, a notable change occurs in the current and transmission spectrum of the sensor. A change in charge density leads to the change in current, which is unique for each nucleotide. Hence, it is possible to sequence the DNA based on the remarkable change in the current (Wasfi et al. 2018).

Recently, an electrochemical biosensor was designed for the alpha-foetoprotein detection for the early diagnosis of liver cancer. Gold nanoparticles (AuNPs) were attached to porous GNR supported on a glassy carbon electrode (GCE). Consequently, anti-AFP was connected to the AuNPs/porous GNR/GCE electrode system. The sensing system showed good electrochemical activity towards the detection of AFP. AuNPs facilitated the electrochemical activity and immobilization of anti-AFP. Similarly, the porous GNR enhanced the active surface area and electrochemical activity (Jothi et al. 2020). An electrochemical label-free immunosensor was fabricated for the accurate and quantitative detection of proprotein convertase subtilisin/kexin type 9 (PCSK9). PCSK 9 is a significant biomarker for cardiovascular disorders. Palladium platinum alloy (PdPt) nanoparticles attached, amino-functionalized fullerene deposited, N-doped GNR attached, GCE (PdPt-fullerene-N-GNR/GCE) was used as the sensing matrix. To prepare a sandwich-type immunosensor, Pt-poly(methylene blue) (Pt-PMB) was combined as a signal label. Meanwhile, a label-free immunosensor was also fabricated with staphylococcus protein A (SPA) immobilized on PdPt-fullerene-N-GNR/GCE. The label-free immunosensor achieved quick detection, and the sandwich-type immunosensor provided ultrasensitive detection of PCSK9 (Li et al. 2017).

A GNR- and silver nanoparticle (AgNP)-based sensitive voltammetric sensor was fabricated to detect histamine accurately. AgNPs and GNRs, supported on an edge plane of pyrolytic graphite (EPPG), were studied with cyclic voltammetry and electrochemical impedance spectroscopy. The proposed sensor exhibited greater sensitivity, selectivity, and recovery rate than a fundamental sample analysis (Kumar and Goyal 2018). Recently, a colourimetric sensor was fabricated to detect dopamine and glutathione by utilizing plasmon hybridization in the hybrid material of AgNP/GNR. Glutathione identification was based on the aggregation of AgNPs induced by glutathione. Aggregation reduces the nanoparticle plasmonic band's absorption intensity and causes a colour change from red to grey. Dopamine detection was based on the etching of nanoparticles, followed by physical changes, making a blue shift in the plasmonic band, and changing colour from green to red (Rostami et al. 2020) (Table 5.2).

5.8.1.2 GNR-based Sensors in the Detection of Drugs

Recently, a sensor was fabricated with a metal-organic framework (HKUST-1) on highly conductive GONRs attached to GCE (HKUST-1/GONR/GCE) to detect the anticancer drug imatinib. Copper hydroxide nanotubes play a significant role in the *in-situ* growth of HKUST-1 on GONRs. The large surface area of the metal-organic framework and GONRs, and the high conductivity of GONRs exhibited a high voltammetric response, two linear ranges, and a shallow limit of detection (Figure 5.4a and b) (Rezvani et al. 2020). Nimesulide is a non-steroidal anti-inflammatory drug (NSAID) used for the treatment of acute pain and inflammatory conditions, but overdosage can cause gastrointestinal and genitourinary problems. Hence, to detect any excess level of nimesulide, an electrochemical sensor was developed with reduced GONRs on an screen printed carbon electrode (SPCE). Compared with

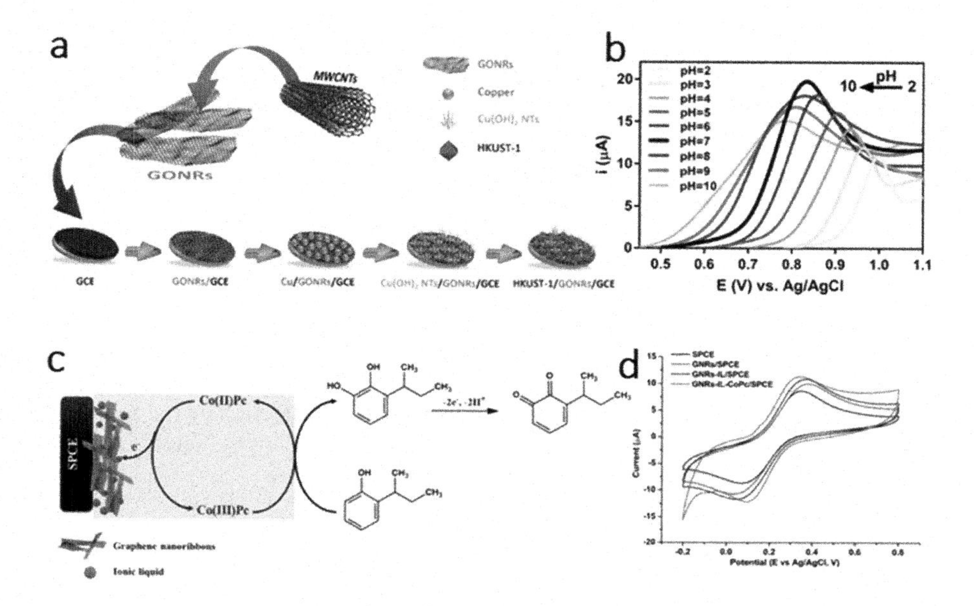

FIGURE 5.4 a) Schematic representation of HKUST-1/GONR/GCE fabrication; b) differential pulse voltammogram of imatinib (0.30 μmol L^{-1} in Britton-Robinson Buffer (BRB) solution) on HKUST-1/GONR/GCE at different pH values. Reprinted with permission from Rezvani Jalal et al. (2020) © 2020 American Chemical Society; c) mechanism of fenubocarb on the electrode; d) cyclic voltammogram of fenubocarb on SPCE, GNR/SPCE, IL/GNR/SPCE, and IL-CoPc/GNR/SPCE. Reprinted with permission from Kalcher and Samphao (2019) © 2019 Elseveir.

MWCNTs, reduced GONRs exhibit good catalytic performance towards the nimesulide because they possess a high surface area, with plenty of functional groups, abundant edge defects, etc. The sensor showed excellent sensitivity in human urine samples and nimesulide tablet samples (Govindasamy et al. 2017b).

5.8.1.3 Detection of Toxic Substances

To prevent insect attack and improve productivity, insecticide and pesticide usage are increasing nowadays. Most of these are harmful to animal and human health. Insecticides, pesticides, harmful chemicals from industry, and other toxic substances can reach humans and animals through food and water. Hence, it is crucial to develop a reliable method to identify harmful chemicals from food and water.

Fenubocarb is a pesticide in the carbamate family which causes severe health hazards to humans and animals. To determine the presence of fenubocarb in food products, a sensor was designed with ionic liquid-cobalt phthalocyanine modified GNRs on an SPCE (IL-CoPc/GNR/SPCE). The amperometric current response was measured from the alkaline hydrolysis product of fenubocarb (2-*sec*-butyl-phenol). The proposed sensor's optimal conditions showed high sensitivity, a wide linear range, a low detection limit, high reproducibility, and repeatability (Figure 5.4c and d) (Kalcher and Samphao 2019). Similarly, to detect the organophosphorus pesticide methyl parathion, a sensor, in which AgNP-decorated GNRs were attached to a SPCE sensor, was designed. The higher electrocatalytic activity of silver and the unique physicochemical characteristics of GNRs combined to exhibit excellent electrochemical activity, high selectivity, and very low detection limits, at nanomolar levels. The sensor showed excellent response towards the methyl parathion detection in real samples of green beans, cabbage, nectarine, and strawberry samples (Govindasamy et al. 2017a).

Bisphenol-A and 4-nonyl-phenol are endocrinal disruptors, for the detection of which separate GNR-based sensors were developed. Compared with graphene, chitosan-modified MWCNTs/GONR hybrid material showed higher electrocatalytic activity towards the detection of bisphenol-A (Xin et al. 2015). A nitrogen-doped GNR-based sensor, modified with molecularly imprinted polymer (MIP) and ionic liquid (IL), was fabricated to detect 4-nonyl-phenol. Detection by the sensor of 4-nonyl-phenol from various water sources showed high sensitivity, selectivity, and stability (Pan et al. 2015).

5.8.2 GNR Applications in Gene Delivery

GNRs supported the anisotropic assembly of DNA base pairs through the non-covalent association of charge transfer interaction and van der Waals forces of attraction to graphene. This donor-acceptor interaction supports the self-assembly of DNA on GNRs. GNR can assemble single-stranded adenine (A), thymine (T), guanine (G), cytosine (C), AT, and GC nucleotides. GNRs act as a double-sided adhesive platform for the attachment of DNA (Reuven et al. 2013).

Graphene-based formulations are regarded as non-viral vectors for gene delivery applications. A non-viral gene delivery agent must have the capacity to load negatively charged nucleic acid, translocate the nucleic acid to cells, and release the nucleic acid to the cytoplasm. GONRs can effectively load up to 6 kilobases (kb) of double-stranded DNA (dsDNA) in a buffer at physiological pH. The study reported that the DNA–GONR conjugation was strong enough even at low ionic strength conditions and in response to treatment with a hydrophobic surfactant. DNA on the GONR platform was expressed efficiently in the nucleus and was found to be safe below 100 μg/mL (Foreman et al. 2017). Up to 100 μg DNA/mL concentration, GONRs did not affect gene expression (Ricci et al. 2017). Loading of single-stranded or double-stranded genetic materials is possible with GONRs without additional functionalization with positively charged materials. Compared with commonly used non-viral vector polyethyleneimine (PEI), the GONR concentration from 20–60 μg/mL showed lower cytotoxicity. GONRs loaded with genetic material exhibited time-dependent enhancement in the delivery of genes at 96–98% transfection efficiency. GONRs underwent intracellular uptake to vesicular structures, with vesicular uptake and entry into the nucleus (Chowdhury et al. 2016).

In a study, the high charge density of PEI and the ultra-high surface area of GNRs combined to achieve the efficient delivery of genetic materials. Compared to PEI or PEI-MWCNTs, PEI-grafted GNRs (PEI-GNR) showed better protection of locked nucleic acid-modified molecular beacon (LNA-modified MB) probes from the digestion of nuclease and interaction with binding proteins. The LNA modified MB/PEI-GNR effectively delivered the gene probe into cells and specifically recognized the targeted microRNA (miRNA) from the cytoplasm with negligible apoptosis and cytotoxicity (Dong et al. 2011).

5.8.3 Application of GNRs in Drug Delivery and Anticancer Activity

The physicochemical nature of GNRs is best suited for the delivery of anticancer agents having aromatic structural properties. Covalent or non-covalent functionalization of nanoribbons with a drug or a targeting moiety can effectively deliver the GNRs to the targeted site for better therapeutic efficacy (Chowdhury et al. 2015b). ZGNRs are functionalized with phenylalanine and doped with boron to make an effective carrier for delivering the drug. Phenylalanine-functionalized, boron-doped (at the edge) ZGNRs exhibit higher chemical reactivity and lower kinetic energy than the functionalized doped ZGNRs in the centre and away from the edge. Increased dipole moment enhances the Gibbs free energy of solvation. The amino group on L-phenylalanine also renders the system hydrophilic. In addition, L-type amino acids are hyper-expressed in tumour environments such as breast cancer. Hence the L-phenylalanine-functionalized boron-doped ZGNRs are promising nanocarriers for tumour-targeted drug delivery applications (Janani and Thiruvadigal 2018a).

The thioxanthene-like antitumour agent lucanthone was loaded onto DSPE-PEG-modified GONRs *via* π-π interactions to treat glioblastoma multiforme by inhibiting over-expressed apurinic endonuclease-1 (Figure 5.5 a). The modification of DSPE-PEG renders the GONR stable in aqueous dispersions. Uptake studies on three different cell lines, U251 (glioblastoma cell line), MCF-7, and CG-4 (both human breast cancer cell lines), showed differential uptake characteristics for DSPE-PEG-GONRs. The apurinic endonuclease-1 over-expressed U251 cell line showed higher uptake than CG-4 and MCF-7 cell lines (Figure 5.5b, c, d, and e). This difference in the uptake of nanocarrier is suitable for the targeted delivery of lucanthone to treat glioblastoma multiforme (Chowdhury et al. 2015a).

Chowdhury et al. studied the uptake characteristics of DSPE-PEG-modified GONRs. Their study suggest that DSPE-PEG-GONR can activate the epidermal growth factor receptors (EGFR), thereby entering the EGFR-overexpressed cells through the macro-pinocytosis-like pathway. Human papillomavirus (HPV) genome-integrated cells also showed greater uptake due to the activated EGFR receptor modulation by the viral protein. Therefore, the study suggests that the DSPE-PEG-GONR system can be used as a carrier for active targeting without the need for additional modification with targeting moieties. Figure 5.6a and b show the mechanism of DSPE-PEG-GONR uptake by EGFR-overexpressed cells and cells integrated with the HPV genome (Chowdhury et al. 2014).

Similarly, in another study, DSPE-PEG-GONR induced differential responses in four different cell lines viz, HeLa cells, NIH-3T3 cells, MCF-7 cells, and SKBR3 cell lines. Viability of all four cell lines was found to be dose- and time-dependent. HeLa cell lines exhibited the highest uptake of all cell lines. Significant cytotoxicity was observed even at the lowest concentrations (10 mg mL^{-1}). MCF-7 and SKBR3

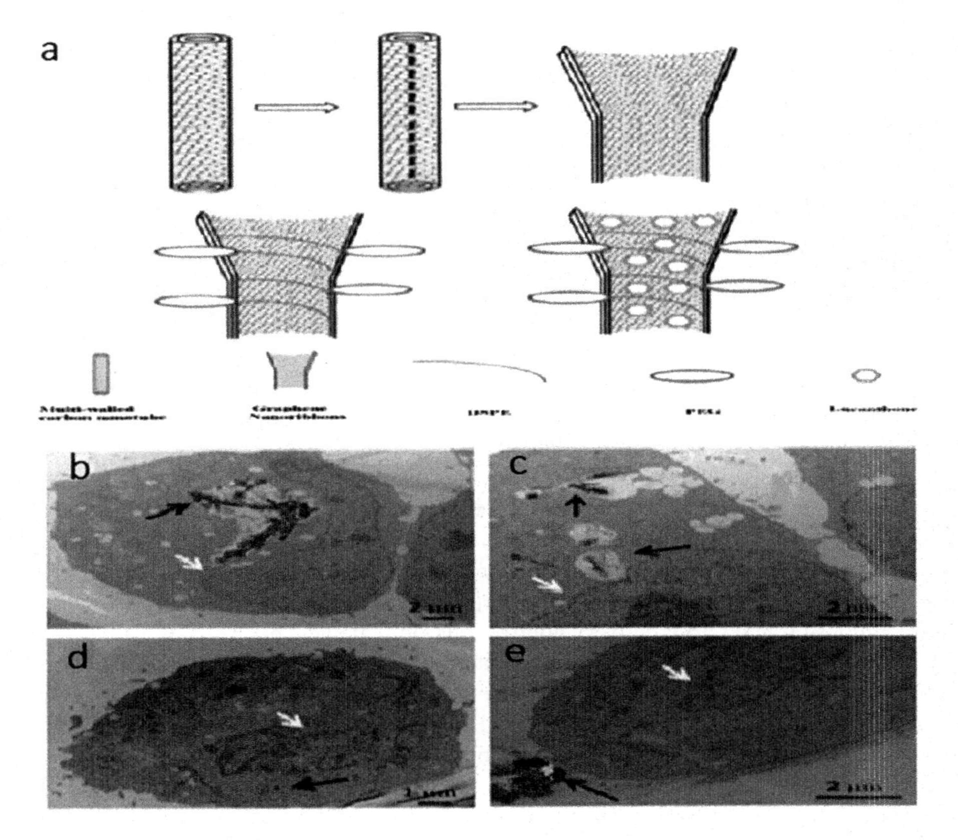

FIGURE 5.5 a) Schematic representation of production of GONR by unzipping of MWCNTs, DSPE-PEG coating, and lucanthone loading. TEM images showing: b) the uptake of large aggregates of DSPE-PEG-GONR into large vesicular structures of CMV/U251; c) uptake of small aggregates of DSPE-PEG-GONR into multiple vesicular structures of CMV/U251; d) significantly fewer uptake of small aggregates by MCF-7 cell lines; e) MCF-7 cells did not show any uptake of large aggregates of DSPE-PEG-GONR. Reprinted with permission from Chowdhury et al. (2015) © 2015 Elsevier.

cells showed very little or no cytotoxicity, and the viability of these cells was 100 % even after 48 hours of incubation with 10 mg mL^{-1} DSPE-PEG-GONR. These reports suggest that DSPE-PEG-GONR exhibited a cell-specific uptake and cytotoxicity (Chowdhury et al. 2013).

In a recent study, the efficiency of GONRs and graphene nanoplatelets (GNPs) for delivering hydrophobic sphingolipid C6 ceramide for anticancer activity was investigated. The shifting of the hydrophobic environment of the ceramide-nanocarrier to the hydrophilic environment allows a large amount of ceramide loading. A 100 μg mL^{-1} concentration of ceramide-loaded GONRs and GNPs reduced the viability of HeLa cells by 93% and 76%, respectively. Live-cell confocal imaging showed that GONRs entered the HeLa cell within 30 min. Compared with GNPs, GONRs were

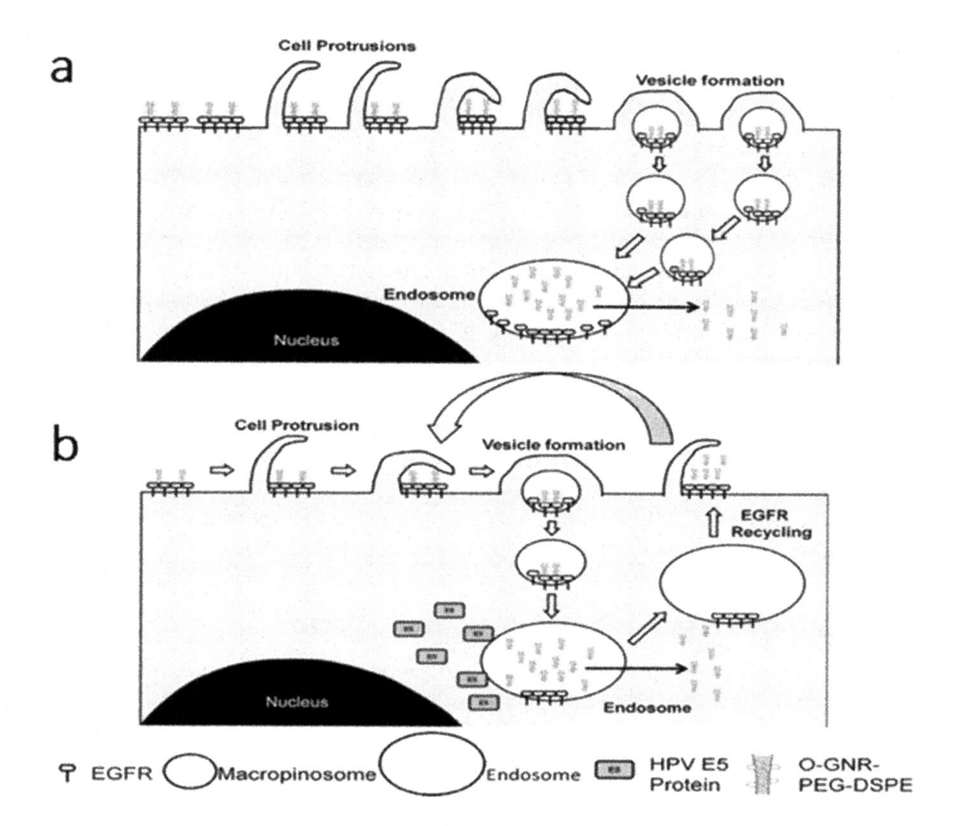

FIGURE 5.6 Schematic illustration of uptake mechanism of EGFR by DSPE-PEG-GONR over-expressed cells. a) Dense EGFR clusters at various locations activated by DSPE-PEG-GONR followed by macro-pinocytic uptake of DSPE-PEG-GONR with receptors; b) Cells with an integrated HPV genome. E5 prevents degradation of activated EGFR receptors, instead recycling them onto the cell surface; this results in a repeated uptake process without further EGFR activation and additional uptake of on nanoparticles on the cell surface or the surrounding regions. Reprinted with permission from Chowdhury, Manepalli, and Sitharaman (2014) © 2014 Elsevier.

the better nanocarrier for the efficient delivery of ceramides to achieve proapoptotic effects (Suhrland et al. 2019). The effect of GONRs at low concentrations in two human neuroblastoma cell lines, SK-N-BE(2) and SH-SY5Y, showed ROS accumulation and the induction of autophagy. However, there was no inhibition of growth or cell death. Hence, GONRs are considered to be a potential carrier of drugs for therapeutic delivery to brain tissues (Mari et al. 2016).

5.8.4 Application of GNRs in Photothermal Therapy

Nanomaterial-based photothermal therapy was developed for the effective treatment of advanced stages of cancer. Nanomaterials exhibiting strong near-IR absorption have been used for effective photothermal therapy. Exposure to photosensitizing agents of cancer cells excited and overheated in the NIR region showed excellent tumour-killing effect and several advantages over surgery, chemotherapy, and radiotherapy(Akhavan et al. 2012).

The use of GONRs, in combination with chemotherapy and photothermal therapy, showed greater tumour-killing effect than conventional chemotherapy. The chemophotothermal efficiency of phospholipid-polyethylene glycol-modified GONRs (PL-PEG-GONRs) loaded with doxorubicin on glioma cell lines (U87) was studied. The synergistic effect of the rise in temperature from laser irradiation and doxorubicin's therapeutic activity reduced the IC_{50} values 6.7-fold compared with free doxorubicin (Lu et al. 2014).

In a study, rGONRs, functionalized with PEG, an arginine-glycine-aspartic acid-based peptide, and cyanine-3, were used for the targeted imaging and photothermal therapy of cancer cells. The PEG-rGONRs exhibited greater NIR absorption than bare GONRs. PEG-rGONRs showed concentration-dependent cytotoxicity and genotoxicity on the U87MG cell line; hence 1 μg mL^{-1} is the optimum concentration for cell imaging and photothermal therapy (Akhavan et al. 2012).

5.8.5 Application of GNRs in Antibacterial Activity

Graphene oxide can be presented as an economical, easy to make, multitarget-specific nanomaterial for preventing bacterial growth. Three possible mechanisms for antimicrobial activity have been observed: 1) sharp-edge penetration by GO and cell membrane damage to a bacterial cell, 2) masking of bacterial cells and inhibition of growth by the basal plane of GO, and 3) ROS production and antibacterial activity (Chen et al. 2014; Akhavan and Ghaderi 2010; Valentini et al. 2019; Liu et al. 2012). The size of the GO sheets also influences the antibacterial activity. Large-sized sheets can induce antibacterial activity by covering and isolating the cells from external nutrients. At the same time, small and defective GO sheets induced more oxidative stress-based inhibition (Liu et al. 2012; Perreault et al. 2015). One study showed that GO could act as an antibiotic and growth promoter at different concentrations. At a low concentration of GO, counterions promote the collision between cells and GO to produce an antibiotic effect, but, at high concentrations, large clusters of GO formed by counterions cause two opposite effects. At low zeta potential, floating scaffolds of GO clusters enhance bacterial growth, whereas, at high zeta potential, scaffolds of GO can inhibit bacterial growth (Palmieri et al. 2017). Very few studies have been reported on the antibacterial activity of GONRs, although mechanism of inhibition similar to those exhibited by GO can also be expected from GONRs.

A study showed that a 100 μg mL^{-1} concentration of GONRs reduced the initial population of *Staphylococcus aureus* and *Escherichia coli* colonies by up to 50%. The exact concentration also exhibited a significant antibiofilm activity towards *S. aureus*. The inhibition of bacterial growth was due to the destruction of the membrane caused by the attraction of electrons from the bacterial membrane (negatively charged) to the edges of GNRs (positively charged) (Ricci et al. 2017). The effect of GONRs on bacterial biofilm activity was studied recently on *Streptococcus mutans*. The antibiofilm effect of GONRs was found to be related to physicochemical properties, such as the presence of oxygen atoms. GONRs can penetrate through the biofilm and reach the bacterial membrane. GONRs exhibited different penetration rates for different organisms (Javanbakht et al. 2020).

5.8.6 Application of GNRs in Bone Regeneration and Locomotor Function Recovery

Regeneration of bone requires a scaffold material with strong osteoinduction and osteogenesis properties. Graphene nanomaterials with specific physicochemical characteristics are potential materials compatible with bone regeneration and are suitable for bone tissue engineering (Hermenean et al. 2016).

GONRs can influence expression of the osteogenic pathway genes and induce mineralization, differentiation, and maturation of osteoblast and the regulation of osteogenesis. A study reported that 200 μg mL^{-1} of GONRs reduced OC, COL-1, and RUNX2 gene expression in the osteogenesis pathway, but 100 μg mL^{-1} GONRs is the safest concentration, did not affect gene expression, and did not induce any cytotoxic effects (Ricci et al. 2017).

The effect of PEGylated GONRs on locomotor function recovery was studied on a rat model. A concentration of 1% GNRs in PEG600 (polyethylene glycol600) (Texas-PEG) was used to recover the locomotor function in the transected spinal cord at lumbar segment level. Texas-PEG has the conducting properties of GNRs and the fusogenic potential of PEG. Fast recovery of the sharply severed spinal cord has been reported 24 hours after surgery (Kim et al. 2016). Further *in-vivo* and *in-vitro* studies on Texas-PEG-induced locomotor recovery reported glial scar formation reduction, neurite regeneration, and fast locomotor function recovery (Kim et al. 2018).

5.8.7 Other Applications

GNRs have been used to enhance the quality of magnetic resonance imaging. Graphene nanomaterials can reduce the toxicity of contrast agents and can improve the quality of the image. Carboxyphenylated GNRs conjugated to aquated gadolinium ions (Gd^{3+}) showed improved contrast of the image compared to gadolinium ions alone (Gizzatov et al. 2014). A free-standing film was fabricated with alginate, chitosan-functionalized GNRs, and graphene flakes for potential use in wound healing and bone tissue engineering. Both the functionalized GNR and graphene flakes films were found to be cytocompatible. Hence, this is a potential biofilm for biomedical applications. Other than biomedical applications, GNRs are used in several fields like energy storage, electronics, engineering, etc. (Silva et al. 2017). Compared with other applications, however, biomedical applications of GNRs are rarely reported.

TABLE 5.1
GNR Functionalization Methods for Various Biomedical Applications

Molecule Used for Modification	Type of Modification	Impact of Modification	Application	References
DSPE-PEG	Non-covalent	Improved aqueous solubility Enhanced EGFR targeting Cell-specific uptake Increased dispersibility	Active targeting without ligand functionalization in tumour therapy	(Chowdhury, Manepalli, and Sitharaman 2014; Chowdhury et al. 2013)
		Increased aromatic drug loading	Delivery of lucanthone to glioblastoma multiforme C6 ceramide for cancer therapy	(Chowdhury et al. 2015b; Suhrland et al. 2019)
		Reduced gene loading efficiency Enhanced stability in buffers and cell culture media	Gene delivery	(Chowdhury et al. 2016)
		Decreased protein binding	—	(Chowdhury et al. 2015)
MnO_2 nanoparticles	Non-covalent	Improved electrocatalytic activity towards hydrogen peroxide	Detection of glucose and hydrogen peroxide	(Vukojević et al. 2018)
Gold nanoparticles	Covalent	Improved electrocatalytic activity	Detection of alpha-foetoprotein	(Jothi, Jaganathan, and Nageswaran 2020)
Gold nanocages	Non-covalent	Improved surface area and conductivity	Detection of target DNA	(Feng et al. 2018)
β-cyclodextrin conjugated graphene sheet	Non-covalent	Enhanced electron transfer	Detection of alpha-foetoprotein	(Li et al. 2015a)
Diazonium salt and surfactant	Covalent and non-covalent	Improved electronic character and solubility	—	Zhu, Higginbotham, and Tour (2009)

(*Continued*)

TABLE 5.1 (CONTINUED)
GNR Functionalization Methods for Various Biomedical Applications

Molecule Used for Modification	Type of Modification	Impact of Modification	Application	References
Graphene sheet	Covalent	Enhanced surface area and conductivity	Detection of ascorbic acid	Lavanya and Gomathi (2015)
CdTe quantum dots	Non-covalent	Amplified electrochemiluminescence	Electrochemiluminescence detection of cholesterol	Huan et al. (2015)
Polyaniline	Covalent	Enhanced electrocatalytic effect	Electrochemical sensing of dobutamine	Asadian et al. (2014)
Metal-organic framework	Non-covalent	Enhanced surface area and conductivity	Electrochemical sensing of imatinib	Rezvani Jalal et al. (2020)
Ionic liquid	Non-covalent	Enhanced electrochemical signal	Electrochemical sensing of fenubocarb	Kalcher and Samphao (2019)
Polyethylene oxide	Covalent	Highest dispersibility in common organic solvents	—	Huang et al. (2016)
L(+)-serine, L(+)-valine and oxygen	Covalent	Modification of electronic character Induced water solubility and biocompatibility	—	Kumar et al. (2019a)
L(+)- phenylalanine and boron-doping	Covalent	The modified structural and electronic character Water solubility	—	Janani and Thiruvadigal (2018a)
L(+)-leucine and group IB elements doping	Covalent	The modified structural and electronic character Water solubility	—	Janani and Thiruvadigal (2017)
Glutamine and minerals	Covalent	Changes in physicochemical characteristics and better aqueous solubility	—	Janani and Thiruvadigal (2018b)

TABLE 5.2
Properties of Various GNR Biosensors

Sensor	Analyte	Sensitivity	Limit of Detection	Linear Range	Applications	References
FAD/apo-GOx/GNR/SPCE	Glucose	—	20 mg L^{-1}	50–2000 mg L^{-1}	Diagnosis of diabetes	Mehmeti et al. (2017)
GNR/GS/Ni hybrid material	Glucose	2.32 mA mM^{-1} cm^{-2}	2.5 nM	5 nM– 5 mM	Diagnosis of diabetes	Lavanya et al. (2017)
Anti-AFP labelled AuNP/porous GNR/GCE	Alpha foetoprotein	—	1 ng mL^{-1}	5–60 ng mL^{-1}	Diagnosis of liver cancer	Jothi, Jaganathan, and Nageswaran (2020)
AgNP/GNR/EPPG	Histamine	0.158 μA $μM^{-1}$	0.049 μM	1–50 μM	Diagnosis of various physiological and pathological conditions	Kumar and Goyal (2018)
Label-free type Sandwich type	PCSK9	—	0.033 ng mL^{-1} 3.33 ng mL^{-1}	0.0001–100 ng mL^{-1} 0.01–100 ng mL^{-1}	Diagnosis of cardiovascular disorders	Li et al. (2017)
AgNP/GNR	Dopamine Glutathione	—	0.04 mM 0.23 μM	0.25–10 μM 1–75 μM	Diagnosis of neurological disorders Diagnosis of cell damage, cancer, leukocyte loss, heart diseases etc.	Rostami et al. (2020)

(Continued)

TABLE 5.2 (CONTINUED)
Properties of Various GNR Biosensors

Sensor	Analyte	Sensitivity	Limit of Detection	Linear Range	Applications	References
HKUST-1/GONR/GCE	Imatinib	45.33 μAL μmol^{-1}	0.006 μM L^{-1}	0.04–1.0 μmol L^{-1} 1–80 μmol L^{-1}	Detection of drug overdose	Rezvani Jalal et al. (2020)
Reduced GONR/SPCE	Nimesulide	1.2 μA μM^{-1} cm^{-2}	3.50 nM	1.0×10^{-8}–1.50×10^{-3} M	Detection of drug overdose	Govindasamy et al. (2017b)
IL-CoPc/GNR/SPCE	Fenubocarb	0.0884 M $A^{-1}.cm^{-2}$	0.0089 μM	0.025–110 μM	Detection of fenubocarb in vegetables	Kalcher and Samphao (2019)
Ag/GNR/SPCE	Methyl parathion	0.5940 μA μM^{-1} cm^{-2}	0.5 nM	0.005–2780 nM	Detection of methyl parathion in vegetables and fruits	Govindasamy et al. (2017a)
Chitosan- modified MWCNTs/ GONR	Bisphenol-A	=	1 ng L^{-1}	0.005–150 μg L^{-1}	Detection of bisphenol-A in water	Xin et al. (2015)
MIP/IL/nitrogen-GNR/GCE	4-Nonyl-phenol	3.4 μA μM^{-1}	8 nM	0.04–6 mM	Detection of 4-nonyl-phenol in water	Pan et al. (2015)

5.9 FUTURE PROSPECTS AND CONCLUSION

GNRs are a new class of emerging nanomaterial offering a wide area of applications. GNRs are applicable to the biomedical field and are also valuable in engineering, energy storage, electronics, etc. However, the biomedical potential of GNRs has not been fully explored. There are very few studies reported on drug delivery applications of GNRs. GNRs can be used for the efficient loading and delivery of highly hydrophobic drug molecules. Bone tissue engineering is the only field of GNR-induced tissue engineering which has been reported, although we can expect the potential applications of GNRs to cardiac, cartilage, skin, and neural tissue engineering in the future. Some reports showed that amino acid-functionalization of GNRs could change the physicochemical characteristics and aqueous solubility of GNRs. Thus, amino acid-functionalized water-soluble GNRs could be used for the development of nano-drug carriers in the future.

The imaging application of GNRs has not yet been well studied. GNRs exhibited high NIR absorbance from the visible to the IR regions, but their applications to photoacoustic and photothermal therapy have not been well explored. Very few studies of GNRs on photothermal therapy have yet been reported. The applications of nanoribbons are entirely based on their physicochemical parameters. There are many methods available for the synthesis of GNRs, but there is still a need to develop an optimum approach to achieve GNRs with exact physicochemical characteristics, high yield, and defect-free nanoribbons. Appropriate biocompatibility and biodegradability studies are required for the in-depth exploration of GNRs for use in the biomedical field.

Various nanoparticles, polymers, genetic materials, antibodies, aptamers, and graphene derivatives can easily conjugate to GNRs. Attachment of multifunctional agents to one platform can develop multimodal GNR-based theranostic platforms. The ease of functionalization of GNRs widens the application of GNRs in the biomedical and other fields. GNR-based investigations are emerging in health care, and incredible combinations of electronic, thermal, mechanical, optical, and edge characteristics make GNRs attractive for research. But the scale-up and commercialization of GNR-based health care products need extensive toxicological studies, standardization of fabrication methods, and some interdisciplinary approaches.

5.10 REFERENCES

Abdolkarimi-Mahabadi, M., and M. Manteghian. 2015. Chemical Oxidation of Multi-Walled Carbon Nanotube by Sodium Hypochlorite for Production of Graphene Oxide Nanosheets.*Fullerenes Nanotubes and Carbon Nanostructures* 23 (10): 860–864.

Afonso, María M., and José Antonio Palenzuela. 2019. *Recent Trends in the Synthesis of Carbon Nanomaterials. Nanomaterials Synthesis.* Elsevier, 2019. 519-555.

Akhavan, Omid, and Elham Ghaderi. 2010. Toxicity of Graphene and Graphene Oxide Nanowalls against Bacteria. *ACS Nano* 4 (10): 5731–5736.

Akhavan, Omid, Elham Ghaderi, and Hamed Emamy. 2012. Nontoxic Concentrations of PEGylated Graphene Nanoribbons for Selective Cancer Cell Imaging and Photothermal Therapy. *Journal of Materials Chemistry* 22 (38): 20626–20633.

Asadian, Elham, Saeed Shahrokhian, Azam Iraji Zad, and Effat Jokar. 2014. In-Situ Electro-Polymerization of Graphene Nanoribbon/Polyaniline Composite Film: Application to Sensitive Electrochemical Detection of Dobutamine. *Sensors and Actuators, B: Chemical* 196: 582–588.

Balarastaghi, Mehran, Vahid Ahmadi, and Ghafar Darvish. 2016. Electro-Optical Properties of New Structure Photodetectors Based on Graphene Nanoribbons: An Ab Initio Study. *Journal of the Optical Society of America B* 33 (11): 2368.

Barzegar, Hamid Reza, Thang Pham, Alexandr V. Talyzin, and Alex Zettl. 2016. Synthesis of Graphene Nanoribbons inside Boron Nitride Nanotubes. *Physica Status Solidi (B) Basic Research* 253 (12): 2377–2379.

Bei, Ho Pan, Yuhe Yang, Qiang Zhang, Yu Tian, Xiaoming Luo, Mo Yang, and Xin Zhao. 2019. Graphene-Based Nanocomposites for Neural Tissue Engineering. In *Molecules*. MDPI AG.

Berahman, M., M. Asad, M. Sanaee, and M. H. Sheikhi. 2015. Optical Properties of Chiral Graphene Nanoribbons: A First Principle Study. *Optical and Quantum Electronics* 47 (10). Springer US: 3289–3300.

Bu, Hao, Yunfei Chen, Min Zou, Hong Yi, Kedong Bi, and Zhonghua Ni. 2009. Atomistic Simulations of Mechanical Properties of Graphene Nanoribbons. *Physics Letters A* 373 (37): 3359–3362.

Campos, Leonardo C., Vitor R. Manfrinato, Javier D. Sanchez-Yamagishi, Jing Kong, and Pablo Jarillo-Herrero. 2009. Anisotropic Etching and Nanoribbon Formation in Single-Layer Graphene. *Nano Letters* 9 (7): 2600–2604.

Casiraghi, C., and D. Prezzi. 2017. *Raman Spectroscopy of Graphene Nanoribbons: A Review*. In: Morandi V., Ottaviano L. (Eds) *GraphITA*. (2017): 19-30 *Carbon Nanostructures*. Springer.

Castellanos-Gomez, Andres, and Bart Jan van Wees. 2013. Band Gap Opening of Graphene by Noncovalent π-π Interaction with Porphyrins. *Scientific Research* 2: 102–108.

Cataldo, Franco, Giuseppe Compagnini, Giacomo Patané, Ornella Ursini, Giancarlo Angelini, Primoz Rebernik Ribic, Giorgio Margaritondo, Antonio Cricenti, Giuseppe Palleschi, and Federica Valentini. 2010. Graphene Nanoribbons Produced by the Oxidative Unzipping of Single-Wall Carbon Nanotubes. *Carbon* 48 (9): 2596–2602.

Celis, A., M. N. Nair, A. Taleb-Ibrahimi, E. H. Conrad, C. Berger, W.A. de Heer, and A. Tejeda. 2016. Graphene Nanoribbons : Fabrication, Properties and Devices. *Journal of Physics. D: Applied Physics* 49: 143001.

Chen, Juanni, Hui Peng, Xiuping Wang, Feng Shao, Zhaodong Yuan, and Heyou Han. 2014. Graphene Oxide Exhibits Broad-Spectrum Antimicrobial Activity against Bacterial Phytopathogens and Fungal Conidia by Intertwining and Membrane Perturbation.*Nanoscale* 6 (3): 1879–1889.

Chen, Qian, Liang Ma, and Jinlan Wang. 2016. Making Graphene Nanoribbons: A Theoretical Exploration. *Wiley Interdisciplinary Reviews: Computational Molecular Science* 6 (3): 243–254.

Chen, Zongping, Wen Zhang, Carlos Andres Palma, Alberto Lodi Rizzini, Bilu Liu, Ahmad Abbas, Nils Richter, et al. 2016. Synthesis of Graphene Nanoribbons by Ambient-Pressure Chemical Vapor Deposition and Device Integration. *Journal of the American Chemical Society* 138 (47): 15488–15496.

Chowdhury, Sayan Mullick, Justin Fang, and Balaji Sitharaman. 2015a. Interaction of Graphene Nanoribbons with Components of the Blood Vascular System. *Future Science OA* 1 (3): fso.15.17.

Chowdhury, Sayan Mullick, Cassandra Surhland, Zina Sanchez, Pankaj Chaudhary, M. A. Suresh Kumar, Stephen Lee, Louis A. Peña, Michael Waring, Balaji Sitharaman, and Mamta Naidu. 2015b. Graphene Nanoribbons as a Drug Delivery Agent for Lucanthone Mediated Therapy of Glioblastoma Multiforme. *Nanomedicine: Nanotechnology, Biology, and Medicine* 11 (1): 109–118.

Corso, Martina, Eduard Carbonell-Sanromà, and Dimas G. de Oteyza. 2018. Bottom-Up Fabrication of Atomically Precise Graphene Nanoribbons. In: De Oteyza D., Rogero C. (Eds) *On-Surface Synthesis II. Advances in Atom and Single Molecule Machines.* Springer: 113-152.

Dimiev, Ayrat M., Artur Khannanov, Iskander Vakhitov, Airat Kiiamov, Ksenia Shukhina, and James M. Tour. 2018. Revisiting the Mechanism of Oxidative Unzipping of Multiwall Carbon Nanotubes to Graphene Nanoribbons. *ACS Nano* 12 (4): 3985–3993.

Dong, Haifeng, Lin Ding, Feng Yan, Hanxu Ji, and Huangxian Ju. 2011. The Use of Polyethylenimine-Grafted Graphene Nanoribbon for Cellular Delivery of Locked Nucleic Acid Modified Molecular Beacon for Recognition of MicroRNA. *Biomaterials* 32 (15): 3875–3882.

Fei, Airong, Qian Liu, Juan Huan, Jing Qian, Xiaoya Dong, Baijing Qiu, Hanping Mao, and Kun Wang. 2015. Label-Free Impedimetric Aptasensor for Detection of Femtomole Level Acetamiprid Using Gold Nanoparticles Decorated Multiwalled Carbon Nanotube-Reduced Graphene Oxide Nanoribbon Composites. *Biosensors and Bioelectronics* 70: 122–129.

Feng, Qiumei, Xiaolei Zhao, Yuehua Guo, Mingkai Liu, and Po Wang. 2018. Stochastic DNA Walker for Electrochemical Biosensing Sensitized with Gold Nanocages@graphene Nanoribbons. *Biosensors and Bioelectronics* 108: 97–102.

Folorunso, Oladipo, Yskandar Hamam, Rotimi Sadiku, Suprakas Sinha Ray, and Gbolahan Joseph Adekoya. 2020. Electrical Resistance Control Model for Polypyrrole-Graphene Nanocomposite: Energy Storage Applications. *Materials Today Communications* 9: 101699.

Foreman, Hui Chen Chang, Gaurav Lalwani, Jaslin Kalra, Laurie T. Krug, and Balaji Sitharaman. 2017. Gene Delivery to Mammalian Cells Using a Graphene Nanoribbon Platform. *Journal of Materials Chemistry B* 5 (12): 2347–2354.

Fujii, Shintaro, and Toshiaki Enoki. 2010. Cutting of Oxidized Graphene into Nanosized Pieces. *Journal of the American Chemical Society* 132 (29): 10034–10041.

Gardener, Jules A., and J. A. Golovchenko. 2012. Ice-Assisted Electron Beam Lithography of Graphene. *Nanotechnology* 23 (18).

Genorio, Bostjan, and Andrej Znidarsic. 2014. Functionalization of Graphene Nanoribbons. *Journal of Physics D: Applied Physics* 47 (9): 094012.

Gizzatov, Ayrat, Vazrik Keshishian, Adem Guven, Ayrat M. Dimiev, Feifei Qu, Raja Muthupillai, Paolo Decuzzi, Robert G. Bryant, James M. Tour, and Lon J. Wilson. 2014. Enhanced MRI Relaxivity of Aquated Gd3+ Ions by Carboxyphenylated Water-Dispersed Graphene Nanoribbons. *Nanoscale* 6 (6): 3059–3063.

Govindasamy, Mani, Veerappan Mani, Shen Ming Chen, Tse Wei Chen, and Ashok Kumar Sundramoorthy. 2017a. Methyl Parathion Detection in Vegetables and Fruits Using Silver@graphene Nanoribbons Nanocomposite Modified Screen Printed Electrode. *Scientific Reports* 7: 1–11.

Govindasamy, Mani, Veerappan Mani, Shen Ming Chen, Thandavarayan Maiyalagan, S. Selvaraj, Tse Wei Chen, Shih Yi Lee, and Wen Han Chang. 2017b. Highly Sensitive Determination of Non-Steroidal Anti-Inflammatory Drug Nimesulide Using Electrochemically Reduced Graphene Oxide Nanoribbons. *RSC Advances* 7 (52). Royal Society of Chemistry: 33043–33051.

Gundra, Kondayya, and Alok Shukla. 2011. Theory of the Electro-Optical Properties of Graphene Nanoribbons. *Physical Review B: Condensed Matter and Materials Physics* 83 (7).

Guo, Zhi-Xin, Dier Zhang, and Xin-Gao Gong. 2009. Thermal Conductivity of Graphene Nanoribbons. *Applied Physics Letters* 95: 163103.

Hermenean, Anca, Sorina Dinescu, Mariana Ionita, and Marieta Costache. 2016. The Impact of Graphene Oxide on Bone Regeneration Therapies. In *Advanced Techniques in Bone Regeneration*. InTech (2016): 151-167.

Higginbotham, Amanda L., Dmitry V. Kosynkin, Alexander Sinitskii, Zhengzong Sun, and James M. Tour. 2010. Lower-Defect Graphene Oxide Nanoribbons from Multiwalled Carbon Nanotubes. *ACS Nano* 4 (4): 2059–2069.

Hou, Zhufeng, and Marcus Yee. 2007. Electronic and Transport Properties of Graphene Nanoribbons. In Proceedings of the 7th IEEE International Conference on Nanotechnology, Hong Kong, China: 554–557.

Hsu, Han, and L. E. Reichl. 2007. Selection Rule for the Optical Absorption of Graphene Nanoribbons. *Physical Review B*, no. 76: 1–5.

Huan, Juan, Qian Liu, Airong Fei, Jing Qian, Xiaoya Dong, Baijing Qiu, Hanping Mao, and Kun Wang. 2015. Amplified Solid-State Electrochemiluminescence Detection of Cholesterol in near-Infrared Range Based on CdTe Quantum Dots Decorated Multiwalled Carbon Nanotubes@reduced Graphene Oxide Nanoribbons. *Biosensors and Bioelectronics* 73: 221–227.

Huang, Yinjuan, Yiyong Mai, Uliana Beser, Joan Teyssandier, Gangamallaiah Velpula, Hans Van Gorp, Lasse Arnt Straasø, et al. 2016. Poly(Ethylene Oxide) Functionalized Graphene Nanoribbons with Excellent Solution Processability. *Journal of the American Chemical Society* 138 (32): 10136–10139.

Jaiswal, Neeraj K., Goran Kovačević, and Branko Pivac. 2015. Reconstructed Graphene Nanoribbon as a Sensor for Nitrogen Based Molecules. *Applied Surface Science* 357: 55–59.

Jaleel, Jumana Abdul, Shabeeba M. Ashraf, Krishnan Rathinasamy, and K. Pramod. 2019. Carbon Dot Festooned and Surface Passivated Graphene-Reinforced Chitosan Construct for Tumor-Targeted Delivery of TNF-α Gene. *International Journal of Biological Macromolecules* 127: 628–636.

Janani, K., and D. John Thiruvadigal. 2017. Chemical Functionalization and Edge Doping of Zigzag Graphene Nanoribbon with L-(+)-Leucine and Group IB Elements: A DFT Study. *Applied Surface Science* 418: 406–413.

Janani, K., and D. John Thiruvadigal. 2018a. Adsorption of Essential Minerals on L-Glutamine Functionalized Zigzag Graphene Nanoribbon-A First Principles DFT Study. *Applied Surface Science* 449: 829–837.

Janani, K., and D. John Thiruvadigal. 2018b. Density Functional Study on Covalent Functionalization of Zigzag Graphene Nanoribbon through L-Phenylalanine and Boron Doping: Effective Nanocarriers in Drug Delivery Applications. *Applied Surface Science* 449: 815–822.

Javanbakht, T., H. Hadian, and K. J. Wilkinson. 2020. Comparative Study of Physicochemical Properties and Antibiofilm Activity of Graphene Oxide Nanoribbons. *Journal of Engineering Science* 7 (1): 1–8.

Jiao, Liying, Xinran Wang, Georgi Diankov, Hailiang Wang, and Hongjie Dai. 2010a. Facile Synthesis of High-Quality Graphene Nanoribbons. *Nature Nanotechnology* 5 (5): 321–325.

Jiao, Liying, Li Zhang, Lei Ding, Jie Liu, and Hongjie Dai. 2010b. Aligned Graphene Nanoribbons and Crossbars from Unzipped Carbon Nanotubes. *Nano Research* 3 (6): 387–394.

Johnson, Asha P., H. V. Gangadharappa, and K. Pramod. 2020. Graphene Nanoribbons: A Promising Nanomaterial for Biomedical Applications. *Journal of Controlled Release* 325: 141–162.

Jothi, Lavanya, Saravana Kumar Jaganathan, and Gomathi Nageswaran. 2020. An Electrodeposited Au Nanoparticle/Porous Graphene Nanoribbon Composite for Electrochemical Detection of Alpha-Fetoprotein.*Materials Chemistry and Physics* 242: 122514.

Kalcher, Kurt, and Anchalee Samphao. 2019. A Highly Sensitive Fenobucarb Electrochemical Sensor Based on Graphene Nanoribbons-Ionic Liquid-Cobalt Phthalocyanine Composites Modified on Screen-Printed Carbon Electrode Coupled with a Flow Injection Analysis. *Journal of Electroanalytical Chemistry* 855: 113630.

Kan, Er-jun, Zhenyu Li, Jinlong Yang, and J. G. Hou. 2008. Half-Metallicity in Edge-Modified Zigzag Graphene Nanoribbons. *Journal of the American Chemical Society* 130 (13): 4224–4225.

Kim, C. Yoon, William K. A. Sikkema, In Kyu Hwang, Hanseul Oh, Un Jeng Kim, Bae Hwan Lee, and James M. Tour. 2016. Spinal Cord Fusion with PEG GNRs (TexasPEG): Neurophysiological Recovery in 24 Hours in Rats. *Surgical Neurology International*: 632–636.

Kim, C. Yoon, William Sikkema, Jin Kim, Jeong Kim, James Walter, Raymond Dieter, Hyung Min Chung, Andrea Mana, James Tour, and Sergio Canavero. 2018. Effect of Graphene Nanoribbons (TexasPEG) on Locomotor Function Recovery in a Rat Model of Lumbar Spinal Cord Transection. *Neural Regeneration Research* 13 (8): 1440–1446.

Kosynkin, Dmitry V., Amanda L. Higginbotham, Alexander Sinitskii, Jay R. Lomeda, Ayrat Dimiev, B. Katherine Price, and James M. Tour. 2009. Longitudinal Unzipping of Carbon Nanotubes to Form Graphene Nanoribbons. *Nature* 458 (7240): 872–876.

Kumar, Janani, Hariharan Rajalakshmi Mohanraj, Preferencial Kala Christian, and John Thiruvadigal David. 2019a. Controlling the Electronic Properties of Zigzag Graphene Nanoribbon Using Amino Acids and Oxygen Molecule-A First Principles DFT Study. *Applied Surface Science* 494: 627–634.

Kumar, Janani, Karthik Peramaiya, Neppolian Bernaurdshaw, and John Thiruvadigal David. 2019b. Functionalization of Zigzag Graphene Nanoribbon with DNA Nucleobases: A DFT Study. *Applied Surface Science* 496: 143667.

Kumar, Neeraj, and Rajendra N. Goyal. 2018. Chemical Silver Nanoparticles Decorated Graphene Nanoribbon Modified Pyrolytic Graphite Sensor for Determination of Histamine. *Sensors & Actuators: B. Chemical* 268: 383–391.

Kumar, Pramod, Pradeep Kumar Singh, Sumit Nagar, Kamal Sharma, and Manish Saraswat. 2020. Effect of Different Concentration of Functionalized Graphene on Charging Time Reduction in Thermal Energy Storage System. *Materials Today: Proceedings* 44 (1): 146–152.

Kumar, Prashant, L. S. Panchakarla, and C. N.R. Rao. 2011. Laser-Induced Unzipping of Carbon Nanotubes to Yield Graphene Nanoribbons. *Nanoscale* 3 (5): 2127–2129.

Kusuma, J., R. Geetha Balakrishna, Siddappa Patil, M. S. Jyothi, H. R. Chandan, and R. Shwetharani. 2018. Exploration of Graphene Oxide Nanoribbons as Excellent Electron Conducting Network for Third Generation Solar Cells. *Solar Energy Materials and Solar Cells* 183: 211–219.

Lavanya, J., and N. Gomathi. 2015. High-Sensitivity Ascorbic Acid Sensor Using Graphene Sheet/Graphene Nanoribbon Hybrid Material as an Enhanced Electrochemical Sensing Platform. *Talanta* 144: 655–661.

Lavanya, J., J. Nithyaa, J. Saravanakumar, and N. Gomathi. 2017. Ultrasensitive and Selective Non-Enzymatic Electrochemical Glucose Sensor Based on Hybrid Material of Graphene Sheet/Graphene Nanoribbon/Nickel Nanoparticle. *Materials Research Bulletin* 98: 300–307.

Li, Jun, Senbin Ye, Tongtao Li, Xinlu Li, Xiaohan Yang, and S. Ding. 2015a. Preparation of Graphene Nanoribbons (GNRs) as an Electronic Component with the Multi-Walled Carbon Nanotubes (MWCNTs). *Procedia Engineering* 102: 492–498.

Li, Na, Hongmin Ma, Wei Cao, Dan Wu, Tao Yan, Bin Du, and Qin Wei. 2015b. Highly Sensitive Electrochemical Immunosensor for the Detection of Alpha Fetoprotein Based on PdNi Nanoparticles and N-Doped Graphene Nanoribbons. *Biosensors and Bioelectronics* 74: 786–791.

Li, Yan, Junlin He, Jun Chen, Yazhen Niu, Yilin Zhao, Yuchan Zhang, and Chao Yu. 2017. A Dual-Type Responsive Electrochemical Immunosensor for Quantitative Detection of PCSK9 Based on n-C60-PdPt/N-GNRs and Pt-Poly (Methylene Blue) Nanocomposites. *Biosensors and Bioelectronic* 101: 7–13.

Liao, Chengzhu, Yuchao Li, and Sie Chin Tjong. 2018. Graphene Nanomaterials: Synthesis, Biocompatibility, and Cytotoxicity. *International Journal of Molecular Sciences* 19 (11): 3564.

Liu, and Speranza. 2019. Functionalization of Carbon Nanomaterials for Biomedical Applications. *C: Journal of Carbon Research* 5 (4): 72.

Liu, Shaobin, Ming Hu, Tingying Helen Zeng, Ran Wu, Rongrong Jiang, Jun Wei, Liang Wang, Jing Kong, and Yuan Chen. 2012. Lateral Dimension-Dependent Antibacterial Activity of Graphene Oxide Sheets. *Langmuir* 28 (33): 12364–12372.

Lu, Yang, and Jing Guo. 2010. Band Gap of Strained Graphene Nanoribbons. *Nano Research* 3 (3): 189–199.

Lu, Yu Jen, Chih Wen Lin, Hung Wei Yang, Kun Ju Lin, Shiaw Pyng Wey, Chia Liang Sun, Kuo Chen Wei, et al. 2014. Biodistribution of PEGylated Graphene Oxide Nanoribbons and Their Application in Cancer Chemo-Photothermal Therapy. *Carbon* 74: 83–95.

Ma, Liang, Jinlan Wang, and Feng Ding. 2012. Recent Progress and Challenges in Graphene Nanoribbon Synthesis. *ChemPhysChem*14(1): 47–54.

Mari, Emanuela, Stefania Mardente, Emanuela Morgante, Marco Tafani, Emanuela Lococo, Flavia Fico, Federica Valentini, and Alessandra Zicari. 2016. Graphene Oxide Nanoribbons Induce Autophagic Vacuoles in Neuroblastoma Cell Lines. *International Journal of Molecular Sciences* 17 (12): 1995.

Marković, Z., S. Jovanović, M. Milosavljević, I. Holclajtner-Antunović, and B. Todorovic-Marković. 2016. Graphene Nanoribbons Synthesis by Gamma Irradiation of Graphene and Unzipping of Multiwall Carbon Nanotubes. In: *Graphene Science Handbook: Fabrication Methods*, 361–374. CRC Press.

Mauri, Emanuele, Aurora Salvati, Antonino Cataldo, Pamela Mozetic, Francesco Basoli, Franca Abbruzzese, Marcella Trombetta, Stefano Bellucci, and Alberto Rainer. 2021. Graphene-Laden Hydrogels: A Strategy for Thermally Triggered Drug Delivery. *Materials Science and Engineering C* 118: 111353.

Mehmeti, Eda, Dalibor M. Stanković, Sudkate Chaiyo, Janez Zavasnik, Kristina Žagar, and Kurt Kalcher. 2017. Wiring of Glucose Oxidase with Graphene Nanoribbons: An Electrochemical Third Generation Glucose Biosensor. *Microchimica Acta* 184 (4): 1127–1134.

Mousavi, Seyyed Mojtaba, Sadaf Soroshnia, Seyyed Alireza Hashemi, Aziz Babapoor, Younes Ghasemi, Amir Savardashtaki, and Ali Mohammad Amani. 2019. Graphene Nano-Ribbon Based High Potential and Efficiency for DNA, Cancer Therapy and Drug Delivery Applications. *Drug Metabolism Reviews* 51 (1): 91–104.

Mullick Chowdhury, Sayan, Gaurav Lalwani, Kevin Zhang, Jeong Y. Yang, Kayla Neville, and Balaji Sitharaman. 2013. Cell Specific Cytotoxicity and Uptake of Graphene Nanoribbons. *Biomaterials* 34 (1): 283–293.

Mullick Chowdhury, Sayan, Prady Manepalli, and Balaji Sitharaman. 2014. Graphene Nanoribbons Elicit Cell Specific Uptake and Delivery via Activation of Epidermal Growth Factor Receptor Enhanced by Human Papillomavirus E5 Protein. *Acta Biomaterialia* 10 (10): 4494–4504.

Mullick Chowdhury, Sayan, Siraat Zafar, Victor Tellez, and Balaji Sitharaman. 2016. Graphene Nanoribbon-Based Platform for Highly Efficacious Nuclear Gene Delivery. *ACS Biomaterials Science and Engineering* 2 (5): 798–808.

Naghdi, Samira, Hyun Yong Song, Alejandro Várez, Kyong Yop Rhee, and Sung Wng Kim. 2020. Engineering the Electrical and Optical Properties of Graphene Oxide via Simultaneous Alkali Metal Doping and Thermal Annealing. *Journal of Materials Research and Technology* 9(6):15824–15837.

Nakano, Hideyuki, Hiroyuki Tetsuka, Michelle J.S. Spencer, and Tetsuya Morishita. 2018. Chemical Modification of Group IV Graphene Analogs. *Science and Technology of Advanced Materials* 19(1): 76–100.

Narita, Akimitsu, Xinliang Feng, Yenny Hernandez, Søren A. Jensen, Mischa Bonn, Huafeng Yang, Ivan A. Verzhbitskiy, et al. 2014. Synthesis of Structurally Well-Defined and Liquid-Phase-Processable Graphene Nanoribbons. *Nature Chemistry* 6 (2): 126–132.

Narita, Akimitsu, Zongping Chen, Qiang Chen, and Klaus Müllen. 2019. Solution and On-Surface Synthesis of Structurally Defined Graphene Nanoribbons as a New Family of Semiconductors.*Chemical Science* 10 (4): 964–975.

Niu, Fang, Zhen Wu Shao, Hong Gao, Li Ming Tao, and Yong Ding. 2021. Si-Doped Graphene Nanosheets for NOx Gas Sensing. *Sensors and Actuators, B: Chemical* 328: 129005.

Orlov, A. V., and I. A. Ovid'ko. 2015. Mechanical Properties of Graphene Nanoribbons: A Selective Review of Computer Simulations. *Reviews on Advanced Materials Science* 40 (3): 249–256.

Ozden, Sehmus, Pedro A.S. Autreto, Chandra Sekhar Tiwary, Suman Khatiwada, Leonardo Machado, Douglas S. Galvao, Robert Vajtai, Enrique V. Barrera, and Pulickel M. Ajayan. 2014. Unzipping Carbon Nanotubes at High Impact. *Nano Letters* 14 (7): 4131–4137.

Palmieri, Valentina, Francesca Bugli, Maria Carmela Lauriola, Margherita Cacaci, Riccardo Torelli, Gabriele Ciasca, Claudio Conti, Maurizio Sanguinetti, Massimiliano Papi, and Marco De Spirito. 2017. Bacteria Meet Graphene: Modulation of Graphene Oxide Nanosheet Interaction with Human Pathogens for Effective Antimicrobial Therapy. *ACS Biomaterials Science and Engineering* 3 (4): 619–627.

Pan, Yanhui, Lei Shang, Faqiong Zhao, and Baizhao Zeng. 2015. A Novel Electrochemical 4-Nonyl-Phenol Sensor Based on Molecularly Imprinted Poly (o-Phenylenediamine-Co-o-Toluidine)-Nitrogen- Doped Graphene Nanoribbons-Ionic Liquid Composite Film. *Electrochimica Acta* 151: 423–428.

Pefkianakis, Eleftherios K., Georgios Sakellariou, and Georgios C. Vougioukalakis. 2015. Chemical Synthesis of Graphene Nanoribbons. *Arkivoc* 2015 (3): 167–192.

Perreault, François, Andreia Fonseca De Faria, Siamak Nejati, and Menachem Elimelech. 2015. Antimicrobial Properties of Graphene Oxide Nanosheets: Why Size Matters. *ACS Nano* 9 (7): 7226–7236.

Pham, Tung, Pankaj Ramnani, Claudia C. Villarreal, Jhoann Lopez, Protik Das, Ilkeun Lee, Mahesh R. Neupane, Youngwoo Rheem, and Ashok Mulchandani. 2019. MoS2-Graphene Heterostructures as Efficient Organic Compounds Sensing 2D Materials. *Carbon* 142: 504–512.

Rabchinskii, M. K., A. S. Varezhnikov, V. V. Sysoev, M. A. Solomatin, Sergei A. Ryzhkov, M. V. Baidakova, D. Yu Stolyarova, et al. 2021. Hole-Matrixed Carbonylated Graphene: Synthesis, Properties, and Highly-Selective Ammonia Gas Sensing. *Carbon* 172: 236–247.

Rajaji, Umamaheswari, Rameshkumar Arumugam, Shen-ming Chen, and Tse-wei Chen. 2018. Graphene Nanoribbons in Electrochemical Sensors and Biosensors : A Review. *International Journal of Electrochemical Science* 13: 6643–6654.

Rastgoo, Morteza, and Morteza Fathipour. 2019. Interaction of DNA Nucleobases with Boron, Nitrogen, and Sulfur Doped Graphene Nano-Ribbon for Sequencing: An Ab Initio Study. *Applied Surface Science* 492: 634–643.

Reina, Giacomo, José Miguel González-Domínguez, Alejandro Criado, Ester Vázquez, Alberto Bianco, and Maurizio Prato. 2017. Promises, Facts and Challenges for Graphene in Biomedical Applications. *Chemical Society Reviews* 46 (15): 4400–4416.

Reuven, Darkeyah G., H. B. Mihiri Shashikala, Sanjay Mandal, Myron N.V. Williams, Jaideep Chaudhary, and Xiao Qian Wang. 2013. Supramolecular Assembly of DNA on Graphene Nanoribbons. *Journal of Materials Chemistry B* 1 (32): 3926–3931.

Rezvani Jalal, Nahid, Tayyebeh Madrakian, Abbas Afkhami, and Arash Ghoorchian. 2020. In Situ Growth of Metal-Organic Framework HKUST-1 on Graphene Oxide Nanoribbons with High Electrochemical Sensing Performance in Imatinib Determination. *ACS Applied Materials and Interfaces* 12 (4): 4859–4869.

Ricci, R., N. C.S. Leite, N. S. da-Silva, C. Pacheco-Soares, R. A. Canevari, F. R. Marciano, T. J. Webster, and A. O. Lobo. 2017. Graphene Oxide Nanoribbons as Nanomaterial for Bone Regeneration: Effects on Cytotoxicity, Gene Expression and Bactericidal Effect. *Materials Science and Engineering C* 78: 341–348.

Rostami, Simindokht, Ali Mehdinia, Ramin Niroumand, and Ali Jabbari. 2020. Enhanced LSPR Performance of Graphene Nanoribbons-Silver Nanoparticles Hybrid as a Colorimetric Sensor for Sequential Detection of Dopamine and Glutathione. *Analytica Chimica Acta* 1120: 11–23.

Saraswat, Vivek, Yuji Yamamoto, Hyun Jung Kim, Robert M. Jacobberger, Katherine R. Jinkins, Austin J. Way, Nathan P. Guisinger, and Michael S. Arnold. 2019. Synthesis of Armchair Graphene Nanoribbons on Germanium-on-Silicon. *Journal of Physical Chemistry C* 123 (30): 18445–18454.

Sharda, Vangmayee, and R. P. Agarwal. 2014. Review of Graphene Nanoribbons. In: 2014 Recent Advances in Engineering and Computational Sciences (RAECS 2014), Chandigarh, India: 6–8.

Shekhirev, Mikhail, and Alexander Sinitskii. 2017. Solution Synthesis of Atomically Precise Graphene Nanoribbons. *Physical Sciences Reviews* 2 (5).

Shinde, Dhanraj B., Joyashish Debgupta, Ajay Kushwaha, Mohammed Aslam, and Vijayamohanan K. Pillai. 2011. Electrochemical Unzipping of Multi-Walled Carbon Nanotubes for Facile Synthesis of High-Quality Graphene Nanoribbons. *Journal of the American Chemical Society* 133 (12): 4168–4171.

Silva, Magda, Sofia G. Caridade, Ana C. Vale, Eunice Cunha, Maria P. Sousa, João F. Mano, Maria C. Paiva, and Natália M. Alves. 2017. Biomedical Films of Graphene Nanoribbons and Nanoflakes with Natural Polymers. *RSC Advances* 7 (44): 27578–27594.

Solís-Fernández, Pablo, Kazuma Yoshida, Yui Ogawa, Masaharu Tsuji, and Hiroki Ago. 2013. Dense Arrays of Highly Aligned Graphene Nanoribbons Produced by Substrate-Controlled Metal-Assisted Etching of Graphene. *Advanced Materials* 25 (45): 6562–6568.

Suhrland, Cassandra, Jean Philip Truman, Lina M. Obeid, and Balaji Sitharaman. 2019. Oxidized Graphene Nanoparticles as a Delivery System for the Pro-Apoptotic Sphingolipid C6 Ceramide. *Journal of Biomedical Materials Research: Part A* 107 (1): 25–37.

Tabarraei, Alireza, Shohreh Shadalou, and Jeong Hoon Song. 2015. Mechanical Properties of Graphene Nanoribbons with Disordered Edges. *Computational Materials Science* 96: 10–19.

Terrones, Mauricio, Andrés R. Botello-méndez, Jessica Campos-delgado, Florentino López-urías, Yadira I. Vega-cantú, Fernando J. Rodríguez-macías, Ana Laura, et al. 2010. Graphene and Graphite Nanoribbons : Morphology, Properties, Synthesis, Defects and Applications. *Nanotoday* 5(4): 351–372.

Tomita, Hiroki, and Jun Nakamura. 2013. Ballistic Phonon Thermal Conductance in Graphene Nanoribbons. *Journal of Vacuum Science & Technology B* 31 (4).

Valentini, Federica, Andrea Calcaterra, Vincenzo Ruggiero, Elena Pichichero, Assunta Martino, Francesca Iosi, Lucia Bertuccini, et al. 2019. Functionalized Graphene Derivatives: Antibacterial Properties and Cytotoxicity. *Journal of Nanomaterials* 2019: 1-14.

Vukojević, Vesna, Sladjana Djurdjić, Miloš Ognjanović, Martin Fabián, Anchalee Samphao, Kurt Kalcher, and Dalibor M. Stanković. 2018. Enzymatic Glucose Biosensor Based on Manganese Dioxide Nanoparticles Decorated on Graphene Nanoribbons. *Journal of Electroanalytical Chemistry* 823: 610–616.

Wakabayashi, Katsunori. 2012. Electronic Properties of Graphene Nanoribbons. *NanoScience and Technology* 57: 277–299.

Wang, Jinlan, Liang Ma, Qinghong Yuan, Liyan Zhu, and Feng Ding. 2011. Transition-Metal-Catalyzed Unzipping of Single-Walled Carbon Nanotubes into Narrow Graphene Nanoribbons at Low Temperature. *Angewandte Chemie: International Edition* 50 (35): 8041–8045.

Wang, Tuo, Zhe Wang, Rodrigo V. Salvatierra, Emily McHugh, and James M. Tour. 2019. Top-down Synthesis of Graphene Nanoribbons Using Different Sources of Carbon Nanotubes. *Carbon* 158: 615–623.

Wasfi, Asma, Falah Awwad, and Ahmad I. Ayesh. 2018. Graphene-Based Nanopore Approaches for DNA Sequencing: A Literature Review. *Biosensors and Bioelectronics* 119: 191–203.

Watanabe, Eiji, Sho Yamaguchi, Jun Nakamura, and Akiko Natori. 2009. Ballistic Thermal Conductance of Electrons in Graphene Ribbons. *Physical Review B: Condensed Matter and Materials Physics* 80 (8): 085404.

Wu, Zhong-Shuai, Wencai Ren, Libo Gao, Bilu Liu, Jinping Zhao, and Hui-Ming Cheng. 2010. Efficient Synthesis of Graphene Nanoribbons Sonochemically Cut from Graphene Sheets. *Nano Research* 3 (1): 16–22.

Xiao, Biwei, Xifei Li, Xia Li, Biqiong Wang, Craig Langford, Ruying Li, and Xueliang Sun. 2014. Graphene Nanoribbons Derived from the Unzipping of Carbon Nanotubes: Controlled Synthesis and Superior Lithium Storage Performance. *Journal of Physical Chemistry C* 118 (2): 881–890.

Xie, Liming, Hailiang Wang, Chuanhong Jin, Xinran Wang, Liying Jiao, Kazu Suenaga, and Hongjie Dai. 2011. Graphene Nanoribbons from Unzipped Carbon Nanotubes: Atomic Structures, Raman Spectroscopy, and Electrical Properties. *Journal of the American Chemical Society* 133 (27): 10394–10397.

Xin, Xiaodong, Shaohua Sun, He Li, Mingquan Wang, and Ruibao Jia. 2015. Electrochemical Bisphenol A Sensor Based on Core – Shell Multiwalled Carbon Nanotubes/Graphene Oxide Nanoribbons. *Sensors and Actuators B: Chemical* 209: 275–280.

Yang, Kai, Liangzhu Feng, Hao Hong, Weibo Cai, and Zhuang Liu. 2013. Preparation and Functionalization of Graphene Nanocomposites for Biomedical Applications. *Nature Protocols* 8 (12): 2392–2403.

Yang, Ming, Lin Weng, Hanxing Zhu, Fan Zhang, Tongxiang Fan, and Di Zhang. 2017. Simultaneously Improving the Mechanical and Electrical Properties of Poly(Vinyl Alcohol) Composites by High-Quality Graphitic Nanoribbons. *Scientific Reports* 7 (1): 1–10.

Yang, Xi, Xiang Yu, and Xin Liu. 2018. Obtaining a Sustainable Competitive Advantage from Patent Information: A Patent Analysis of the Graphene Industry. *Sustainability* 10 (12).

Yazyev, Oleg V. 2013. A Guide to the Design of Electronic Properties of Graphene Nanoribbons. *Accounts of Chemical Research* 46 (10): 2319–2328.

Zakaria, A. B. M., and Danuta Leszczynska. 2019. Electrochemically Prepared Unzipped Single Walled Carbon Nanotubes-MnO2 Nanostructure Composites for Hydrogen Peroxide and Glucose Sensing. *Chemosensors* 7 (1).

Zhang, Baomei, Yang Wang, and Guangxi Zhai. 2016. Biomedical Applications of the Graphene-Based Materials. *Materials Science and Engineering C* 61: 953–964.

Zhao, Pengcheng, Meijun Ni, Yiting Xu, Chenxi Wang, Chao Chen, Xiurui Zhang, Chunyan Li, Yixi Xie, and Junjie Fei. 2019. A Novel Ultrasensitive Electrochemical Quercetin Sensor Based on MoS2 - Carbon Nanotube @ Graphene Oxide Nanoribbons / HS-Cyclodextrin/Graphene Quantum Dots Composite Film. *Sensors and Actuators, B: Chemical* 299: 126997.

Zhu, Yu, Amanda L. Higginbotham, and James M. Tour. 2009. Covalent Functionalization of Surfactant-Wrapped Graphene Nanoribbons. *Chemistry of Materials* 21 (21): 5284–5291.

6 Biobased Nano Materials (Plant-based for Green Materials) Synthesis, Properties and Their Application in Biomedical Science

Anita Tilwari and Rajesh Saxena

CONTENTS

DOI: 10.1201/9781003110781-6

6.1 INTRODUCTION

Nanotechnology is an essential branch of science, intersecting with many other branches, and has generated extensive research in recent years, impacting on all life forms. Nanotechnology deals with materials with particle sizes in the nanometre scale (one dimension less than 100 nm). The first idea of the term nanotechnology emerged from a lecture delivered by Richard Feynman in 1959. The prefix "nano" is a Greek word that signifies "dwarf".

Silver nanoparticles (AgNPs) have been widely studied. Nanotechnology, based on consequent nanostructures, was discovered quite recently, because nano silver-based materials proved to have qualities that are extremely intriguing, challenging, and promising, and are appropriate for different biomedical applications, particularly in improving customized medical care practice. Silver nanoparticles have a high capability for developing beneficial antimicrobial proxies, drug-delivery, analysis and detection systems, biomaterials and clinical instrument coatings, tissue healing, complicated health care strategies, and overall improved performance for better therapeutic options. Many studies have been carried out to understand the mechanistic pathway of the biological interactions and possible toxic effects associated with nanoparticles.

Cutting-edge innovations and clear understanding have cleared a path through the barriers blocking the use of biological substances, for example, plants and microorganisms, in the blend of nanoparticles and innovative work in natural and restorative plant science towards the development of nanotechnology. The prospects of using plants and their extracts or components are a step towards creating nanoparticles and are an essential source for the dependable and ecologically protective technique of synthesizing metallic-nanoparticles and their characterization. Nanoparticles based on plants are synthesized mainly using shoots, roots, extracts, barks, leaves of medicinal plants, shrubs, trees, and microbe-mediated nanoparticles, with the help of algae, bacteria, fungi, and actinomycetes, etc. Plants contain compounds of beneficial therapeutic value and have been exploited since ancient times as conventional or herbal medicines. Because of their considerable variety, plants have been investigated continually for a broad range of uses, and unlicenced collection of plants from the wild from species at risk might represent a danger to the plant realm. Hence, biotemplate synthesis has vast potential to show some impact in coming years, in which nanoparticles based on plant-based biomolecules can be synthesized in a rapid, one-step protocol that can overcome the mentioned disadvantages of collecting wild plant material, contributing towards green and bio- principled nanoparticle production.

The bottom-up approach is followed for bio-based nanoparticle synthesis, and plant proteins, secondary metabolites, extracts, etc., act as stabilizing agents. In comparison with some other chemical methods, this method helps to synthesize more stable and uniform nanoparticles. These nanoparticles have shown high catalytic activity, fluorescence activity, ion-exchange capacity, and greater surface area, and are present in different forms, such as metals, ceramics, and magnetic forms, which allow nanoparticles and their applications in the agriculture field (like soil nutrients, environmental decontamination, crop protection, pollutant detection, post-harvest waste reduction, growth stimulation, and the controlled delivery of nutrients, fertilizers, genetic materials and pesticides, and analysis of soil structure (Panpatte et al. 2016)), as well as medicine, biosensor development, water purification, food packaging, cosmetic industries, personal care products, and delivery systems of therapeutic agents for better medical treatments (Morin-Crini et al., 2019, Alsammarraie et al., 2018).

Nanoparticles are regarded as a cross-over between bulk size materials and atomic structures. Furthermore, because of their small size and substantial surface/mass ratio, inorganic nanoparticles are remarkable. Gold (Au) and silver (Ag) have added much consideration due to their unique role in various scientific arenas like catalysis, biosensing, and optics. Specifically, silver nanoparticles (AgNPs) have been used as a catalyst for accelerating some chemical reactions due to their antimicrobial and antioxidant properties (Alsammarraie et al., 2018). Reports on plants and microorganisms show that nanoparticle production has benefits like being quick, readily accessible, safe to deal with, and involving a considerable range of biomolecules to intercede in nanoparticle synthesis.

6.2 MECHANISMS OF BIOSYNTHESIS OF NANOPARTICLES

Different procedures can be used for synthesizing nanoparticles, namely chemical, physical, or biological methods. In chemical methods, nanoparticle synthesis processes constitute a significant concern, especially the use of hazardous chemicals like capping and stabilizing agents, making these processes environmentally non-protective. This represents a major challenge to be overcome for nanoparticles to be used in clinical settings, requiring major steps to develop a reliable, biocompatible, non-toxic, eco-friendly, and green approach for synthesizing NPs and their formulations for different applications (Korbekandi et al., 2009, 2012, Castro et al., 2013, Iravani and Zolfaghari, 2013). This can be achieved by using alternate biological approaches where natural resources, such as enzymes, biodegradable polymers, microorganisms, vitamins, polysaccharides, and biological systems are involved. One of the approaches having immense potential for controlled and modular methods for producing monodispersed stable NPs is contingent on the biosynthesis of NPs by fungi, bacteria, and yeasts (bottom-up method) instead of physical or chemical approaches (Iravani et al., 2011). To this end, a number of species of plant (leaf, seed, fruit, bark, etc.), microorganisms (isolates, extracts, or whole cells of algae, bacteria, fungi, etc.), and animal sources (Mukherjee et al., 2017), have been used for the development of green nanotechnology. Bacteria and plants are considered to be potential bio-factories for synthesizing low-cost, novel, eco-friendly, stable, and monodispersed nanoparticles like gold, titanium, cadmium sulphide, silver, platinum, titanium dioxide, palladium, magnetite, etc. These biosynthesized NPs include a green chemistry approach using biological sources, excluding the involvement of toxic chemicals for synthesis, and are beneficial in terms of operating expenses, production, and disposal of reagents and wastes. These biosynthesized NPs have high dependability, exhibit upgraded biocompatibility, and practically negligible cytotoxicity, with the differet properties when synthesized in different ways.

Researchers use biomass or cell extracts of micro-organisms for nanoparticle synthesis (e.g. magnetotactic bacteria and bacterial S-layers). *Trichoderma asperellum*, a viable, fungal biocontrol agent against plant pathogens, was used for the first time to synthesize nano-crystalline silver nanoparticles (sizes in the range 13–18 nm) (Mukherjee et al., 2008). Many fungus- and bacterium- (cell-free extracts and biomass) mediated Ag, Au and titanium dioxide nanoparticles were synthesized (Chinnaperumal et al., 2018, Cuevas et al., 2015, Fayaz et al., 2010). The therapeutic potential of the *in-situ* surface coating of these biosynthesised NPs is attributed to phytochemicals or effective chemical components through the reduction process. The size, shape, and biological composition of the NPs boost the potential biological and medical applications of these NPs.

6.3 BIOLOGICAL SYNTHESIS OF NANOPARTICLES BY PLANTS

Nanomedicine involves nanomaterials with applications in the biological and clinical fields, having a significant impact on unique innovations and advanced processes of disease therapy, sensing, and diagnosis (Rizzo et al., 2013, Ovais et al., 2018, Mukherjee et al., 2016). Recently, scientists worldwide have been working on the fabrication and design of such nanomaterials and their investigations to identify desired therapeutic and biological effects. Aimed at different biomedical applications, such as photothermal treatment, drug delivery, nanocomposites, medical imaging, etc., many nanoparticles like polymer nanoparticles (Elsabahy and Wooley, 2012), quantum dots (Fang et al., 2012), carbon nanotubes (Mocan et al., 2017), dendrimers (Svenson and Tomalia, 2006), and liposomes (Bozzuto and Molinari, 2015), as well as inorganic metal nanoparticles (Barui et al., 2017, Nethi et al., 2019), have been established. The nanoparticles have shown excellent applications in biomedicine, which is attributed to the nanoscale characteristics of the material as compared with the bulk scale.

An alternative technique to synthetic chemical approaches, biological synthesis is better, more cost-effective, non-hazardous, and does not require toxic chemicals.

The process of biological synthesis of nanomaterials includes selection of the organism species, optimized settings for cell growth, genetic properties of the organism, and the reduction reaction, which is supervened by the nucleation process, in which the transitional metal nuclei function as a template for the crystal development, capping and neutralization of the biosynthesized nanoparticles (Singh and Kundu, 2014). Microbial cultures are advantageous due to low-cost production, high growth rate, and easily manageable growth conditions, like temperature, pH, oxygenation, and incubation time (Ovais et al., 2018). Some examples of NPs produced using microbial resources are gold (Au) (Chakravarty et al., 2015), iron (Fe) (Sundaram et al., 2012), silver (Ag) (Shivaji et al., 2011), zinc (Zn) (Jayaseelan et al., 2012), magnetite (Fe_3O_4) (Elblbesy et al., 2014), and cadmium (Cd) (Tripathi et al., 2014).

6.3.1 GOLD NANOPARTICLES (AuNPs)

Gold nanoparticles have remarkable physicochemical properties, a small size, simple synthesis, surface modification, and minimal toxicity, which have been utilized for a wide range of biomedical applications, for example, photothermal treatment, nucleic acid delivery, bio-imaging, drug delivery, biosensors and so on. (Balakrishnan et al., 2017, Mukherjee et al., 2016, Dykman and Khlebtsov, 2012).

6.3.2 Silver Nanoparticles (AgNPs)

These can also be readily synthesized and modified in size and shape, and show enormous potential in biomedical applications, especially in anticancer, antibacterial, antifungal, and bio-sensing, as well as agricultural applications (Patra et al., 2018, Mukherjee et al., 2014, 2017).

6.3.3 Graphene Quantum Dots (GQDs)

GQDs are graphene blocks with 2-D (two-dimensional) transverse sizes less than 100 nm and excellent biological properties. Biomass waste, which is green, cheap, abundant, readily available, and rich in carbon, can be considered a potential precursor for synthesising GQDs. Leaves of green plants can be used to produce graphene. Luminous GQD nanoparticles, mainly semiconductor-based, including CdS, ZnS, and PbS nanoparticles, have been synthesized by bacteria and plants, and have potential applications in cell imaging and cell labelling, or as a fluorescent biomarker (Iravani et al., 2014).

6.3.4 Magnetite Nanoparticles

Magnetite nanoparticles are iron oxide nanoparticles, and, due to their valuable properties, like synthesis route, low cost, biocompatibility, and application in the biomedical field, have gained much attention (Veiseh et al., 2010). Several other nanoparticles have been synthesized utilizing microbes, including bacteria, with nanoparticles like platinum, copper, cobalt, nickel, palladium, zinc oxide and titanium oxide NPs being used mainly. Spherical TiNPs (40 to 60 nm diameter) were synthesized extracellularly by bacterial cultures with the help of *Lactobacillus* sp. at room temperature (Prasad et al., 2007). Zinc oxide nanoparticles have been used in many fields, including antibacterial, biomedical, and agricultural applications (Barui et al., 2017).

6.4 BIOSYNTHESIS OF NPS BY MICROORGANISMS TAKES PLACE BY TWO DIFFERENT MECHANISMS

6.4.1 Bioreduction

A standard and easy method for microbe facilitated synthesis of NPs is bioreduction. This procedure gives a steady supply of dispersed crystalline NPs, which can be simply separated from other debris. Where the metal ions are initially trapped inside or on the surface, they are afterwards reduced into stable nanoparticle forms with the help of micro-organisms or their proteins, as well as by enzymic oxidation (Deplanche et al., 2010). Nanoparticle formation occurs proceeding over the surface of the fungal mycelia by electrostatic contact, assuming that silver or gold ions have been buried at first into the surface of the fungal cells within the slightly negatively charged cell wall and ions from the carboxylic groups of the enzymes (Li et al., 2011).

6.4.2 Biosorption

The bisorption NP biosynthesis mechanism involves the uptake of precursor metal ions into the microorganism, by the creation of organic polymers, giving rise to mineral formation that can affect nucleation by favouring the modification of complete minerals.

Kalishwaralal and coworkers synthesized the AgNPs in the bacterium *Bacillus licheniformis* through the nitrate reductase enzyme, or by the enzymatic electron transport metal reduction process, with nitrate ions and silver ions reduced to metallic silver. *B. licheniformis* reduces Ag^+ to Ag^0, with the subsequent production of silver nanoparticles, by secreting an essential cofactor for NADH-dependent enzymes, like nitrate reductase (Li et al., 2011).

Metallophilic microorganisms can form the foundation for the production of metallic nanoparticles. These organisms flourish under high concentrations of various heavy metal ions, such as the efflux streams from metal processing plants, mining waste piles, and natural mineralized zones (Patel et al., 2020), and contain various metal tolerance gene clusters that enable detoxification of cells, using several mechanisms, such as metal efflux, complexation, or reductive precipitation. To regulate metal homoeostasis and achieve tolerance to the toxic environment, some metal ions, for example, Ag^+, Co^{2+}, Hg^{2+}, Cu^{2+}, Cd^{2+}, Ni^{2+}, Zn^{2+}, and Pb^{2+} induce a transcriptomic and proteomic response in the microorganism (Emamverdian et al., 2015).

Bacterial magnetic particles (BacMPs) are synthesized, regulating their morphology and size, from magnetotactic bacteria. BacMPs are enclosed with a thin organic membrane, making them ideal biotechnological material, superior to artificial magnetites. Biomineralization of BacMPs has a multistep synthesis route (Arakaki et al., 2008). Primarily, the vesicle formed by enfolding acts as the precursor of the organic membrane. However, the envelope-forming mechanism is unclear. The mechanism for creating vesicles in magnetotactic bacteria is comparable to that of most eukaryotes, in that a specific GTPase facilitates the act of enfolding. The vesicles subsequently form a chain laterally with cytoskeletal filaments. In the next phase, BacMP biomineralization, ferrous ions enter into vesicles with the help of transmembrane iron receptors. Transport proteins and siderophores internalize the exogenous iron. The oxidation-reduction system controls the internal iron concentration. In the final step, BacMP proteins stimulate magnetite crystal nucleation as well as modulating their structure. Different proteins that are integrated into the BacMP membrane show significant roles in magnetite generation, including the buildup of supersaturating iron concentrations, iron oxidation to encourage mineralization, preservation of reducing conditions, or the incomplete reduction and desiccation of ferrihydrite to magnetite (Arakaki et al., 2008). BacMPs are used in immunoassays, cell separation, and drug screening. In addition, completely robotized systems for achieving single nucleotide polymorphism segregation and DNA recovery frameworks have been created to utilize such functionalized BacMPs. Nano-sized magnetic elements have great potential in developing novel nanotechniques.

Another possible mechanism consisting of both active and passive mechanisms for synthesising magnetites, using the bacterium *Shewanella oneidensis*, was suggested by Perez-Gonzalez and coworkers in their recent study (Ali et al., 2019). Firstly, bacteria use ferrihydrite as a terminal acceptor of electrons, resulting in the active production of Fe^{2+}, and changes in the pH value cause an upsurge in the bacterial breakdown of amino acids. Subsequently, the concentration of Fe^{2+} and Fe^{3+} at the negatively charged cell wall and cell fragments causes a local upsurge of supersaturation of the magnetite arrangement and precipitates the magnetite phase (Trivedi and Bergi, 2021).

6.5 APPLICATIONS OF PLANT-MEDIATED NANOPARTICLES

Nanomedicine is a rapidly growing research area for the detection, diagnosis, and treatment of human illnesses (Boulaiz et al., 2011). Detached nanoparticles are typically employed in nanomedicine as fluorescent biological tags (Wolfbeis, 2015), gene and drug delivery mediators (Suri et al., 2007) as well as in applications such as pathogen detection (Mocan et al., 2017), tissue engineering (Giustini et al., 2010, Chatterjee et al., 2011), treatment of tumours *via* heating (hyperthermia) (Chang et al., 2018), MRI contrast improvement, and phagokinetic studies (De La Isla et al., 2003).

Many research and review papers have been published on the uses of nanoparticles in the biomedicine field (Salata, 2004). Studies on biosynthesized nanoparticles discovered their usefulness in a broad range of applications, such as drug delivery, DNA analysis, gene therapy, antibacterial, antiviral, and anticancer activities, MRI, biosensors, and biocatalysis.

The delivery of vaccines and drugs to the target cell is challenging; here, nanoparticles display a significant role. New strategies, such as nanocarriers, have overcome the restrictions of the conventional modes of delivery of drugs and vaccines. Combining drugs or vaccines with the nanocarriers can provide an operative approach for delivering the drug/vaccine to the target site for treating several diseases like malignance, autoimmune diseases, infections, neurobiological diseases, and many more.

6.5.1 Application in Vaccine Delivery against Coronaviruses (COVs)

The mRNA-based vaccines mRNA-1273, BNT162a1, b1, b2 and c2 (BioNTech) and LUNAR-COV19 (Arcturus Therapeutics), used for COVID-19 treatment by targeting the S-layer protein and other specific regions, are a combination of mRNA vaccine constituent with a nanocarrier. Similarly, nanocarriers can be used to deliver antigens, evade early degradation of these particles in the body, and support translating these particles into efficient immunogens, sidestepping potential side effects triggered by the treatment (Pati et al., 2018).

In COVs, the RNA/DNA vaccines developed using the NPs are delivered through a different method; firstly the NPb-Vs method, in which the RNA/DNA or the antigen is combined with the nanocarrier, and secondly, the antigen is attached to the surface of nanocarrier, exposing it to the surroundings. The nanocarrier encapsulates the antigen or DNA/RNA vaccine to defend antigens against proteolytic degradation or allow the vaccine to be directly targeted towards antigen-presenting cells (APCs) (Heinrich et al., 2020). The APCs start the NPb-V (nanoparticle based vaccine) and direct the antigen near the surface of the cell or translate the mRNA into the particular antigen. In this process, the NPB-V is indirectly aimed at delivering the "cargo" to the APCs but to imitate the virus itself.

6.5.2 Drug Delivery

Delivering the drugs specifically and securely at the precise time to the targeted sites with controlled delivery to accomplish the maximum therapeutic impact is central to the development and advancement of novel drug delivery methods. Targeted nanocarriers essentially circumnavigate the fences of blood-carrying vessels to reach target cells. Drug vectors effectively enter the target cells by crossing cell membranes through endocytotic and transcytotic transport mechanisms (Yetisgin et al., 2020).

As nanoparticle drug carriers are very small, they can pass through the blood-brain barrier and the restricted epithelial intersections that generally obstruct the transport of drugs towards the target location. Moreover, because of their higher surface area/volume ratio, nanocarriers display better biodistribution and pharmacokinetics of the therapeutic agents carried, whereas minimization of toxicity happens due to accumulation only at the target site (Rizvi and Saleh, 2018). Modifications in the solubility of hydrophobic mixtures are made to achieve appropriate concentrations of the therapeutic agent at the target site without distal accumulation, while the stability of the therapeutic agent is increased by the delivery system (Chenthamara et al., 2019).

Magnetic nanoparticles like magnetite (Fe_3O_4) or maghemite (Fe_2O_3) appear to be biocompatible, and have been used for the heat treatment of cancer (magnetic hyperthermia), steered drug delivery, gene therapy, stem cell cataloging as well as MRI and DNA analysis (Ganapathe et al. 2020). Xiang et al. (2007) gauged the cytotoxicity to *in vitro* mouse fibroblasts of magnetosomes from *Magnetospirillum gryphiswaldense*. Their study observed that the decontaminated and pasteurized magnetosomes were not harmful to mouse fibroblasts *in vitro*. A recent study indicated the impact of bacterial magnetic elements on the mouse immune response (Meng et al., 2010), where ovalbumin was used as an antigen in the experiment, with Freund's adjuvant, phosphate buffer, and BacMPs being tested to protect BALB/c mice. Fourteen days later, anti-ovalbumin titres (IgG) and subtype (IgG1, IgG2), the propagation capability of T lymphocytes, as well as the appearance of IFN-gamma, IL-10, IL-4, and IL-2 were measured. The results revealed the non-influence of the mouse immune response by the native bacterial magnetic particles (BMPs), so that magnetosomes can be used as a novel carrier of drugs or genes for tumour treatment. Another study (Sun et al., 2011) suggested that the anticancer agent doxorubicin (DOX), loaded on bacterial magnetosomes (BMs), inhibited tumour growth *via* covalent bonding. This research was carried out on H22 mice possessing tumours, and the DOX-loaded BMs displayed a similar tumour reduction rate similar to that achieved alone (86.8% *vs* 78.6%, respectively), but causing much lower cardiac risk. Though the particles were administrated intravenously to the hard tumour in this fundamental research, this could possibly control the delivery of these drug-stacked BMs, causing them to aggregate and accomplish their therapeutic effects next to the tumours.

With respect to the pharmacokinetics and biocompatibility of BMs (Sun et al., 2009), the study monitored BM distribution in excreta, serum, urine, and key organs following BM injection into the sublingual vena of Sprague-Dawley rats. Specific methods were implemented to sterilize BMs and to use BMs with a narrow size range, and BMs were found only in livers, with no evidence of BMs in excreta or urine in the 72 h following intravenous injection.

Magnetotactic bacteria (MTB) MC-1, having the unique organelles, magnetosomes, is an extensively utilized mediator for drug transport. The route of each MTB, containing a mixture of nanoparticles, flagella, and magnetite, was steered in small-sized blood vessels (Felfoul, 2011) by functional magnetotaxis. It was proven that the magnetosomes in every MTB could be used to track the movement of these bacteria by using an MRI system, because such magnetosomes interrupt the nearby magnetic field, disturbing the relaxation time throughout MRI. However, to manually direct these MTBs closer to a target, it is crucial to image these living bacteria *in vivo* using a scientific imaging modality. Magnetic resonance weighted images and the relativity of MTB were studied to confirm the ability to supervise and direct MTB drug delivery procedures, using a scientific scanner. It was observed that MTB influenced the relaxation time significantly, suggesting the concept of a negative divergence agent. As the signal decay in the weighted images were translated into bacterial concentration, a detection curve of weighted images changed proportionally to the weighted image.

Xie et al. (2009) described their work using MTB-NPs for gene transport. They directed the use of PEI (polyethyleneimine)-related MTB-NPs to supply β-galactosidase plasmids both *in vivo* and *in vitro*. They concluded that such MTB-PEI-NP setups are particularly green as they are far less toxic than PEI alone.

Gold and its compounds have been utilized as therapeutic agents since the commencement of human civilization, with the earliest record tracing back 5000 years in Egypt (Pan et al., 2014). AuNPs possess a specific shape and size, which contribute to their optical and electrical properties, along with a high surface/volume ratio; the surface of AuNPs can also be modified through the use of ligands containing functional groups such as thiols, amines, and phosphines, which show compatibility for gold surfaces (Giljohann et al., 2010). Gold nanoparticles emerged as a promising framework for delivering drugs and genes, providing a valuable complement to conventional delivery systems. The mixture of minimal cytotoxicity, more surface vicinity, steadiness, and feature tunability provides specific attributes that enable new delivery techniques. Chemically synthesized AuNPs had earlier been studied for biomedical applications, however, there have been no reviews of the use of biosynthesized AuNPs for drug transport to the best of our knowledge.

Kalishwaralal et al. (2009) observed that silver nanoparticles delivered by *B.s licheniformis* exhibited anti-angiogenic activity. Silver nanoparticles have been widely being used as antifungal, antibacterial, anti-inflammatory, and antiviral agents. Bovine retinal endothelial cells (BRECs) had been treated with diverse groupings of silver nanoparticles for 24 h in the presence or absence of vascular endothelial growth factor (VEGF), with a 500 nM (IC_{50}) silver nanoparticle suspension being capable of inhibiting the expansion and spread of BRECs. The cells displayed a marked increase in caspase-3 activity with respect to DNA ladders, a marker of programmed cell death or apoptosis. The results showed that Ag nanoparticles hinder BREC BRECcell survival using the PI3K/Akt-dependent pathway.

It is expected that nanoparticle-mediated directed drug delivery may fundamentally decrease the concentrations of anticancer drugs used, with greater specificity, increased survival, and negligible toxicity. We believe that we will see a rise in nanotechnology-based diagnostic and therapeutic facilities within the next few years. Furthermore, customized medicine is another area where nanotechnology can play a vital role. Because of the heterogeneity of malignancy and the development of anticancer drug resistance, a specific, directed treatment with good effectiveness is not possible. In addition, magnetic nanoparticles might be utilized for hyperthermal treatment of cancers. Hyperthermal cancer treatment includes directing magnetic nanoparticles inside the body, explicitly targeting the sites of malignant tissue. Localized heating at exact positions is enabled by the help of an external magnetic field (De Jong et al., 2008).

A novel drug delivery system, based on arginine-glycine-aspartic acid (RGD)-conjugated graphene quantum dots (GQDs), was synthesized and used to direct the antitumour drug doxorubicin (DOX) to achieve targeted cancer fluorescence imaging as well as tracking and monitoring of drug delivery without the need for external dyes. The inherently stable fluorescence of GQDs enables real-time monitoring of the cellular uptake of the DOX-GQDs-RGD nano-assembly and the consequent release of DOX. The release of DOX demonstrated strong pH dependence, implying a hydrogen-bonding interaction between GQDs and DOX, an observation that suggests the possibility of GQDs serving as pH-sensitive drug carriers. As a nanocarrier, GQDs unlinked with DOX were non-toxic to U251 human glioma cells. Compared to free DOX, DOX-GQDs-RGD conjugates demonstrated substantial cytotoxicity to U251 glioma cells over a broad range of DOX concentrations. After applying the GQDs as drug carriers, the drug efficacy of DOX improved without concomitant increases in DOX dosage. Cellular uptake results of DOX-GQDs-RGD conjugates indicate that not only DOX but also some GQDs penetrated cell nuclei after 16 h of incubation. This enhancement, combined with efficient nuclear delivery, improved the cytotoxicity of DOX dramatically. This type of drug delivery system, based on GQDs, may find widespread applications in biomedicine.

6.5.3 Antibacterial Agents

With the evolution and increased frequency of pathogenic micro-organisms resistant to antibiotics, Ag-based antibacterial agents have been emphasized in recent years. Silver nanoparticles have been biosynthesized using the fungus *Trichoderma viride* (Elgorban et al., 2016). The concentration of aqueous silver ions (Ag^+) in solution decreased, when exposed to a filtrate of *T. viride*, forming very stable silver nanoparticles (AgNPs) within the size range 5–40 nm and were also demonstrated to improve antimicrobial activities, in the presence of various antibiotics, against both Gram-negative and Gram-positive bacteria. The antibacterial activities of kanamycin, ampicillin, chloramphenicol, and erythromycin increased in the presence of AgNPs, compared with antibiotic preparations without AgNPs, with the maximum effect being observed for ampicillin. Thus, the results provided helpful insights into the development of new antimicrobial agents, with antibiotic and AgNP combinations showing greater antimicrobial effects. Alghuthaymi et al. (2015) showed that silver nanoparticles, produced extracellularly with the help of the fungus *Fusarium oxysporum* embedded within textile fabrics, decreased infection by human pathogenic bacteria like *Staphylococcus aureus*.

6.5.4 Biosensors

Nanoparticles can be used in various systems, including biosensors. Round Se-nanoparticles, moulded by *Bacillus subtilis* and having diameters of 50–400 nm were reported (Wang et al., 2010). These circular monoclinic Se nanoparticles can be transformed into extraordinarily anisotropic, 1-D, trigonal structures at ambient temperature after 24 h after synthesis. Moreover, selenium nanomaterial crystals, with a high surface/volume ratio, high biocompatibility, and suitable cohesive potential, have been employed as an improving and stable substance for constructing a biosensor for horseradish peroxidase (HRP). Because Se nanomaterials exhibit reasonable cohesive capacity and bioactivity, they exhibit moderate electrocatalytic activity close to decreasing concentrations of H_2O_2 (hydrogen peroxide). These H_2O_2 (hydrogen peroxide) biosensors exhibited hypersensitivity and affinity for H_2O_2 with a detection level of i-th detection limit activity and affinity of nanomaterials.

Moreover, nanoparticle biosensors have detected hydrogen peroxide in meal, medical, pharmaceutical, industrial, and environmental samples. Zheng et al. (2010) demonstrated that AuAg alloy nanoparticles, biosynthesized using yeast cells, had been employed to make a sensitive electrochemical vanillin sensor. Electrochemical analysis showed that the vanillin sensor, primarily constructed of an AuAg alloy nanoparticle-modified glassy carbon electrode, boosted the electrochemical response to vanillin approximately five fold. Below the optimum operating conditions, the oxidation peak current of vanillin on the sensor was directly correlated with its concentration in the range 0.2–50 μM, with a low detection limit of 40 nM. This vanillin sensor efficiently analyzed vanilla bean and tea samples to determine vanillin, signifying that it could have real-world applications in vanillin-monitoring systems. In additional research, AuNP-based glucose oxidase (GO_x) biosensors have been developed, so that AuNPs can increase the enzyme activity of GO_x (Lipińska et al., 2021). The inline response range of the glucose biosensor is 20 μM–0.80 mM glucose, with a 17 μM detection limit. This biosensor could be used to determine the glucose concentration in commercial glucose injections.

6.5.5 Biocatalytic Agents

Nanoparticles have been widely used to enhance reactions as reductants or catalysts because they have high surface areas and particular traits (Khan et al., 2019). To improve microbiological reaction rates, several types of magnetic nanoparticles have been used. Shan et al. (2005) utilized the bacterial cells of *Pseudomonas delafieldii* with the help of magnetic Fe_3O_4 nanoparticles to achieve dibenzothiophene desulphurization. Magnetic nanoparticles were the handiest nanoparticles not only because of their catalytic features but also for their great capacity to diffuse. The very high surface energy of nanoparticles enhance the adsorption at the cells. After applying a magnetic field externally, the bacterial cells became well distributed in the solution even without mixing, which increased the opportunity for collecting cells for reprocessing. The consequences were that the desulphurization efficacy of *P. delafieldii* have not been brought down, and the cells can be recycled in numerous instances. To brush up microbiological rate of reaction, some magnetic nanoparticles have been used. Shan et al. (2005) used coated microbial cells of *Pseudomonas delafieldii* with magnetic Fe_3O_4 nanoparticles to achieve dibenzothiophene desulfurization. Magnetic nanoparticles have been utilized for their catalytic function and their excellent capability to dissolve. Because of high surface energies, strong adsorption by nanoparticles can be seen on the cells. After applying an external magnetic field, the cells were diffused nicely in the solution irrespective of mixing, which allowed the collection of cells for regeneration. The outcomes displayed the desulfurization efficacies of *Ps. delafieldii* barely decreased, and the cells could be regenerated many more times.

6.5.6 Magnetic Separation and Detection

The combination of biological molecules and magnetic particles makes an attractive material for constructing assay systems. Viable chemiluminescence enzyme-linked immunochemical assay using immobilized antibodies on BacMPs has been developed for sensitive detection of microscopic molecules, like pollutants from the environment and hormones (Cinquanta et al., 2017, Abdulsattar and Greenway, 2019). Xenoestrogens, such as bisphenol A (BPA), alkylphenol ethoxylates, and linear alkylbenzene sulfophates (LAS), have been detectable by the use of monoclonal antibodies restrained on BacMPs, primarily based on the aggressive reaction of xenoestrogens. The time required to complete the entire process was around fifteen minutes, whereas the conventional methods might take more than 2.5 hours. But this process provided a more comprehensive detection assay and a lower detection limit, compared with ELISA, where, for comparison, the same type of antibodies were used.

Surface-tuning of magnetic nanoparticles is a particularly exciting area of research, having numerous applications. The surfaces of BacMPs may be adapted using aminosilane compounds for DNA extraction using magnetic nanoparticle systems. A solid-phase adsorbent (magnetic particles) is appropriate for DNA extraction techniques as it can be manipulated through simple methods.

6.6 CONCLUSION

Nanotechnology has tremendous applications in the chemical, physical, and biomedical fields. This chapter describes the mechanism of biosynthesis by plants of nanoparticles, such as plants and micro-organisms. For example, gold NPs, silver NPs, GQDs and magnetite NPs, by different methods. The biosynthesis of nanoparticles by micro-organisms by other synthetic methods, such as bioreduction and biosorption, is also discussed in detail. Nanotechnology has vast biomedical applications, such as in the form of targeted drug carriers, as well as antibacterial agents. It is expected that nanoparticle-mediated directed drug delivery systems may fundamentally decrease the amounts of anticancer drugs being used, with greater accuracy and specificity, feasibility, and negligible toxicity. Nanoparticles can also be used as a biocatalytic agent for increasing reaction rates. Nanoparticles have been widely used to alter the speed of reactions as reductants or catalysts, because of their huge surface areas and particular traits. Biosensors and magnetic separation and detection are other potential uses of nanotechnology, combinations of biological molecules and magnetic nanoparticles making an attractive material for building advanced assay systems.

6.7 REFERENCES

Abdulsattar J.O., Greenway G.M. A sensitive chemiluminescence based immunoassay for the detection of cortisol and cortisone as stress biomarkers. *Journal of Analytical Science and Technology*. 2019 Dec;10(1):1–3.

Alghuthaymi M.A., Almoammar H., Rai M., Said-Galiev E., Abd-Elsalam K.A. Myconanoparticles: synthesis and their role in phytopathogens management. *Biotechnology & Biotechnological Equipment*. 2015 Mar 4;29(2):221–36.

Ali J., Ali N., Wang L., Waseem H., Pan G. Revisiting the mechanistic pathways for bacterial mediated synthesis of noble metal nanoparticles. *Journal of Microbiological Methods*. 2019 Apr 1;159:18–25.

Alsammarraie F.K., Wang W., Zhou P., Mustapha A., Lin M. Green synthesis of silver nanoparticles using turmeric extracts and investigation of their antibacterial activities. *Colloids and Surfaces B: Biointerfaces*. 2018 Nov 1;171:398–405.

Arakaki A., Nakazawa H., Nemoto M., Mori T., Matsunaga T. Formation of magnetite by bacteria and its application. *Journal of the Royal Society Interface*. 2008 Sep 6;5(26):977–99.

Balakrishnan S., Mukherjee S., Das S., Bhat FA., Raja Singh P., Patra C.R., Arunakaran J. Gold nanoparticles–conjugated quercetin induces apoptosis via inhibition of EGFR/PI3K/Akt–mediated pathway in breast cancer cell lines (MCF-7 and MDA-MB-231). *Cell Biochemistry and Function*. 2017 Jun;35(4):217–31.

Barui A.K., Kotcherlakota R., Bollu V.S., Nethi S.K., Patra C.R. Biomedical and drug delivery applications of functionalized inorganic nanomaterials. In *Biopolymer-Based Composites* 2017 Jan 1 (pp. 325–379). Woodhead Publishing, Sawston, UK.

Becerra-Castro C., Kidd P.S., Rodríguez-Garrido B., Monterroso C., Santos-Ucha P., Prieto-Fernández Á. Phytoremediation of hexachlorocyclohexane (HCH)-contaminated soils using Cytisus striatus and bacterial inoculants in soils with distinct organic matter content. *Environmental Pollution*. 2013 Jul 1;178:202–10.

Boulaiz H., Alvarez P.J., Ramirez A., Marchal J.A., Prados J., Rodríguez-Serrano F., Perán M., Melguizo C., Aranega A. Nanomedicine: application areas and development prospects. *International Journal of Molecular Sciences*. 2011 May;12(5):3303–21.

Bozzuto G., Molinari A. Liposomes as nanomedical devices. *International Journal of Nanomedicine*. 2015;10:975.

Chakravarty I., Pradeepam R.J., Kundu K., Singh P.K., Kundu S. Mycofabrication of gold nanoparticles and evaluation of their antioxidant activities. *Current Pharmaceutical Biotechnology*. 2015 Aug 1;16(8):747–55.

Chang D., Lim M., Goos J.A., Qiao R., Ng Y.Y., Mansfeld F.M., Jackson M., Davis T.P., Kavallaris M. Biologically targeted magnetic hyperthermia: potential and limitations. *Frontiers in Pharmacology*. 2018 Aug 2;9:831.

Chatterjee D.K., Diagaradjane P., Krishnan S. Nanoparticle-mediated hyperthermia in cancer therapy. *Therapeutic Delivery*. 2011 Aug;2(8):1001–14.

Chenthamara D., Subramaniam S., Ramakrishnan S.G., Krishnaswamy S., Essa M.M., Lin F.H., Qoronfleh M.W. Therapeutic efficacy of nanoparticles and routes of administration. *Biomaterials Research*. 2019 Dec;23(1):1–29.

Chinnaperumal K., Govindasamy B., Paramasivam D., Dilipkumar A., Dhayalan A., Vadivel A., Sengodan K., Pachiappan P. Bio-pesticidal effects of Trichoderma viride formulated titanium dioxide nanoparticle and their physiological and biochemical changes on Helicoverpa armigera (Hub.). *Pesticide Biochemistry and Physiology*. 2018 Jul 1;149:26–36.

Cinquanta L., Fontana D.E., Bizzaro N. Chemiluminescent immunoassay technology: what does it change in autoantibody detection?. *Autoimmunity Highlights*. 2017 Dec;8(1):1–8.

Cuevas J.M., Geller R., Garijo R., López-Aldeguer J., Sanjuán R. Extremely high mutation rate of HIV-1 in vivo. *PLoS Biology*. 2015 Sep 16;13(9):e1002251.

De Jong W.H., Borm P.J. Drug delivery and nanoparticles: applications and hazards. *International Journal of Nanomedicine*. 2008 Jun;3(2):133.

De La Isla A., Brostow W., Bujard B., Estevez M., Rodriguez J.R., Vargas S., Castano V.M. Nanohybrid scratch resistant coatings for teeth and bone viscoelasticity manifested in tribology. *Materials Research Innovations*. 2003 Apr 1;7(2):110–4.

Deplanche K., Caldelari I., Mikheenko I.P., Sargent F., Macaskie L.E. Involvement of hydrogenases in the formation of highly catalytic Pd (0) nanoparticles by bioreduction of Pd (II) using Escherichia coli mutant strains. *Microbiology*. 2010 Sep 1;156(9):2630–40.

Dykman L., Khlebtsov N. Gold nanoparticles in biomedical applications: recent advances and perspectives. *Chemical Society Reviews*. 2012;41(6):2256–82.

Elblbesy M.A., Madbouly A.K., Hamdan T.A. Bio-synthesis of magnetite nanoparticles by bacteria. *American Journal of Nano Research and Applications*. 2014;2(5):98–103.

Elgorban A.M., Al-Rahmah A.N., Sayed S.R., Hirad A., Mostafa A.A., Bahkali A.H. Antimicrobial activity and green synthesis of silver nanoparticles using Trichoderma viride. *Biotechnology & Biotechnological Equipment*. 2016 Mar 3;30(2):299–304.

Elsabahy M., Wooley K.L. Design of polymeric nanoparticles for biomedical delivery applications. *Chemical Society Reviews*. 2012;41(7):2545–61.

Emamverdian A., Ding Y., Mokhberdoran F., Xie Y. Heavy metal stress and some mechanisms of plant defense response. *The Scientific World Journal*. 2015 Oct;2015(4).

Fang M., Peng C.W., Pang D.W., Li Y. Quantum dots for cancer research: current status, remaining issues, and future perspectives. *Cancer Biology & Medicine*. 2012 Sep;9(3):151.

Fayaz A.M., Balaji K., Girilal M., Yadav R., Kalaichelvan P.T., Venketesan R. Biogenic synthesis of silver nanoparticles and their synergistic effect with antibiotics: a study against gram-positive and gram-negative bacteria. *Nanomedicine: Nanotechnology, Biology and Medicine*. 2010 Feb 1;6(1):103–9.

Felfoul O. *MRI-based Tumour Targeting Enhancement with Magnetotactic Bacterial Carriers* 2011. Ecole Polytechnique, Montreal (Canada).

Ganapathe L.S., Mohamed M.A., Mohamad Yunus R., Berhanuddin D.D. Magnetite (Fe3O4) nanoparticles in biomedical application: from synthesis to surface functionalisation. *Magnetochemistry*. 2020 Dec;6(4):68.

Giljohann D.A., Seferos D.S., Daniel W.L., Massich M.D., Patel P.C., Mirkin C.A. Gold nanoparticles for biology and medicine. *Angewandte Chemie International Edition.* 2010 Apr 26;49(19):3280–94.

Giustini A.J., Petryk A.A., Cassim S.M., Tate J.A., Baker I., Hoopes P.J. Magnetic nanoparticle hyperthermia in cancer treatment. *Nano Life.* 2010 Mar 1;1(01n02):17–32.

Heinrich M.A., Martina B., Prakash J. Nanomedicine strategies to target coronavirus. *Nano Today.* 2020 Aug 31:100961.

Iravani S. Green synthesis of metal nanoparticles using plants. *Green Chemistry.* 2011;13(10):2638–50.

Iravani S., Zolfaghari B. Green synthesis of silver nanoparticles using Pinus eldarica bark extract. *BioMed Research International.* 2013 Oct;2013:639725.

Iravani S., Korbekandi H., Mirmohammadi S.V., Zolfaghari B. Synthesis of silver nanoparticles: chemical, physical and biological methods. *Research in Pharmaceutical Sciences.* 2014 Nov;9(6):385.

Jayaseelan C., Rahuman A.A., Kirthi A.V., Marimuthu S., Santhoshkumar T., Bagavan A., Gaurav K., Karthik L., Rao K.B. Novel microbial route to synthesize ZnO nanoparticles using Aeromonas hydrophila and their activity against pathogenic bacteria and fungi. *Spectrochimica Acta Part A: Molecular and Biomolecular Spectroscopy.* 2012 May 1;90:78–84.

Kalishwaralal K., Banumathi E., Pandian S.R., Deepak V., Muniyandi J., Eom S.H., Gurunathan S. Silver nanoparticles inhibit VEGF induced cell proliferation and migration in bovine retinal endothelial cells. *Colloids and Surfaces B: Biointerfaces.* 2009 Oct 1;73(1):51–7.

Khan I., Saeed K., Khan I. Nanoparticles: properties, applications and toxicities. *Arabian Journal of Chemistry.* 2019 Nov 1;12(7):908–31.

Korbekandi H., Iravani S., Abbasi S. Production of nanoparticles using organisms. *Critical Reviews in Biotechnology.* 2009 Dec 1;29(4):279–306.

Korbekandi H., Iravani S., Abbasi S. Optimization of biological synthesis of silver nanoparticles using Lactobacillus casei subsp. casei. *Journal of Chemical Technology & Biotechnology.* 2012 Jul;87(7):932–7.

Lee K.X., Shameli K., Yew Y.P., Teow S.Y., Jahangirian H., Rafiee-Moghaddam R., Webster T.J. Recent developments in the facile bio-synthesis of gold nanoparticles (AuNPs) and their biomedical applications. *International Journal of Nanomedicine.* 2020;15:275.

Li X., Xu H., Chen Z.S., Chen G. Biosynthesis of nanoparticles by microorganisms and their applications. *Journal of Nanomaterials.* 2011 Oct;2011: 1687–4110.

Lipińska W., Grochowska K., Siuzdak K. Enzyme immobilization on gold nanoparticles for electrochemical glucose biosensors. *Nanomaterials.* 2021 May;11(5):1156.

Meng C., Tian J., Li Y., Zheng S. Influence of native bacterial magnetic particles on mouse immune response. *Wei sheng wu xue bao= Acta Microbiologica Sinica.* 2010 Jun 1;50(6):817–21.

Mocan T., Matea C.T., Pop T., Mosteanu O., Buzoianu A.D., Suciu S., Puia C., Zdrehus C., Iancu C., Mocan L. Carbon nanotubes as anti-bacterial agents. *Cellular and Molecular Life Sciences.* 2017 Oct 1;74(19):3467–79.

Mocan T., Matea C.T., Pop T., Mosteanu O., Buzoianu A.D, Puia C., Iancu C., Mocan L. Development of nanoparticle-based optical sensors for pathogenic bacterial detection. *Journal of Nanobiotechnology.* 2017 Dec;15(1):1–4.

Morin-Crini N., Lichtfouse E., Torri G., Crini G. Applications of chitosan in food, pharmaceuticals, medicine, cosmetics, agriculture, textiles, pulp and paper, biotechnology, and environmental chemistry. *Environmental Chemistry Letters.* 2019 Dec;17(4):1667–92.

Mukherjee A., Dutta D., Banerjee S., Ringø E., Breines E.M., Hareide E., Chandra G., Ghosh K. Potential probiotics from Indian major carp, Cirrhinus mrigala. Characterization, pathogen inhibitory activity, partial characterization of bacteriocin and production of exoenzymes. *Research in Veterinary Science*. 2016 Oct 1;108:76–84.

Mukherjee P., Roy M., Mandal B.P., Dey G.K., Mukherjee P.K., Ghatak J, Tyagi A.K., Kale SP. Green synthesis of highly stabilized nanocrystalline silver particles by a non-pathogenic and agriculturally important fungus T. asperellum. *Nanotechnology*. 2008 Jan 29;19(7):075103.

Mukherjee S., Chowdhury D., Kotcherlakota R., Patra S. Potential theranostics application of bio-synthesized silver nanoparticles (4-in-1 system). *Theranostics*. 2014;4(3):316.

Mukherjee, S., Patra, C.R., 2017. Biologically synthesized metal nanoparticles: recent advancement and future perspectives in cancer theranostics. *Future science OA* 3.3(2017):FSO203.

Mukherjee S., Sahu P., Halder G. Microbial remediation of fluoride-contaminated water via a novel bacterium Providencia vermicola (KX926492). *Journal of Environmental Management*. 2017 Dec 15;204:413–23.

Nethi S.K., Das S., Patra C.R., Mukherjee S. Recent advances in inorganic nanomaterials for wound-healing applications. *Biomaterials Science*. 2019;7(7):2652–74.

Ovais M., Khalil A.T., Ayaz M., Ahmad I., Nethi S.K., Mukherjee S. Biosynthesis of metal nanoparticles via microbial enzymes: a mechanistic approach. *International Journal of Molecular Sciences*. 2018 Dec;19(12):4100.

Pan S.Y., Litscher G., Gao S.H., Zhou S.F., Yu Z.L., Chen H.Q., Zhang S.F., Tang M.K., Sun J.N., Ko K.M. Historical perspective of traditional indigenous medical practices: the current renaissance and conservation of herbal resources. *Evidence-Based Complementary and Alternative Medicine*. 2014 Apr 27;2014:525340.

Panpatte DG, Jhala YK, Shelat HN, Vyas RV. Nanoparticles: the next-generation technology for sustainable agriculture. In *Microbial Inoculants in Sustainable Agricultural Productivity* 2016 (pp. 289–300). Springer, New Delhi.

Patel A., Enman J., Gulkova A., Guntoro P.I., Dutkiewicz A., Ghorbani Y., Rova U, Christakopoulos P, Matsakas L. Integrating biometallurgical recovery of metals with biogenic synthesis of nanoparticles. *Chemosphere*. 2020 Sep 14:128306.

Pati R., Shevtsov M., Sonawane A. Nanoparticle vaccines against infectious diseases. *Frontiers in Immunology*. 2018 Oct 4;9:2224

Patra J.K., Das G., Fraceto L.F., Campos E.V., del Pilar Rodriguez-Torres M., Acosta-Torres L.S., Diaz-Torres L.A., Grillo R., Swamy M.K., Sharma S., Habtemariam S. Nano based drug delivery systems: recent developments and future prospects. *Journal of Nanobiotechnology*. 2018 Dec;16(1):1–33.

Prasad N.K., Rathinasamy K., Panda D., Bahadur D. Mechanism of cell death induced by magnetic hyperthermia with nanoparticles of γ-Mn x Fe 2–x O 3 synthesized by a single step process. *Journal of Materials Chemistry*. 2007;17(48):5042–51.

Rizvi S.A., Saleh A.M. Applications of nanoparticle systems in drug delivery technology. *Saudi Pharmaceutical Journal*. 2018 Jan 1;26(1):64–70.

Rizzo L.Y., Theek B., Storm G., Kiessling F., Lammers T. Recent progress in nanomedicine: therapeutic, diagnostic and theranostic applications. *Current Opinion in Biotechnology*. 2013 Dec 1;24(6):1159–66.

Salata O.V. Applications of nanoparticles in biology and medicine. *Journal of Nanobiotechnology*. 2004 Dec;2(1):1–6.

Shan G., Xing J., Zhang H., Liu H. Biodesulfurization of dibenzothiophene by microbial cells coated with magnetite nanoparticles. *Applied and Environmental Microbiology*. 2005 Aug;71(8):4497–502.

Shivaji S., Madhu S., Singh S. Extracellular synthesis of antibacterial silver nanoparticles using psychrophilic bacteria. *Process Biochemistry*. 2011 Sep 1;46(9):1800–7.

Singh P.K., Kundu S. Biosynthesis of gold nanoparticles using bacteria. *Proceedings of the National Academy of Sciences, India Section B: Biological Sciences*. 2014 Jun 1;84(2):331–6.

Sun J., Li Y., Liang X.J., Wang P.C. Bacterial magnetosome: a novel biogenetic magnetic targeted drug carrier with potential multifunctions. *Journal of Nanomaterials*. 2011 Jan 1;2011:469031–469043.

Sun J.B., Wang Z.L., Duan J.H., Ren J., Yang X.D., Dai S.L., Li Y. Targeted distribution of bacterial magnetosomes isolated from Magnetospirillum gryphiswaldense MSR-1 in healthy Sprague-Dawley rats. *Journal of Nanoscience and Nanotechnology*. 2009 Mar 1;9(3):1881–5.

Sundaram P.A., Augustine R., Kannan M. Extracellular biosynthesis of iron oxide nanoparticles by Bacillus subtilis strains isolated from rhizosphere soil. *Biotechnology and Bioprocess Engineering*. 2012 Aug 1;17(4):835–40.

Suri S.S., Fenniri H., Singh B. Nanotechnology-based drug delivery systems. *Journal of Occupational Medicine and Toxicology*. 2007 Dec;2(1):1–6.

Svenson S., Tomalia D.A. Dendrimers as nanoparticulate drug carriers. In *Nanoparticulates as Drug Carriers* 2006 (p. 298). Imperial College Press, London.

Tripathi S., Srirambalaji R., Singh N., Anantharaman G. Chiral and achiral helical coordination polymers of zinc and cadmium from achiral 2, 6-bis (imidazol-1-yl) pyridine: Solvent effect and spontaneous resolution. *Journal of Chemical Sciences*. 2014 Sep 1;126(5):1423–31.

Trivedi R., Bergi J. Application of bionanoparticles in wastewater treatment. In *Advanced Oxidation Processes for Effluent Treatment Plants* 2021 Jan 1 (pp. 177–197). Elsevier, Amsterdam, the Netherlands.

Veiseh O., Gunn J.W., Zhang M. Design and fabrication of magnetic nanoparticles for targeted drug delivery and imaging. *Advanced Drug Delivery Reviews*. 2010 Mar 8;62(3):284–304.

Wang T., Yang L., Zhang B., Liu J. Extracellular biosynthesis and transformation of selenium nanoparticles and application in H2O2 biosensor. *Colloids and Surfaces B: Biointerfaces*. 2010 Oct 1;80(1):94–102.

Wolfbeis O.S. An overview of nanoparticles commonly used in fluorescent bioimaging. *Chemical Society Reviews*. 2015;44(14):4743–68.

Xiang L., Wei J., Jianbo S., Guili W., Feng G., Ying L. Purified and sterilized magnetosomes from Magnetospirillum gryphiswaldense MSR-1 were not toxic to mouse fibroblasts in vitro. *Letters in Applied Microbiology*. 2007 Jul;45(1):75–81.

Xie J., Chen K., Chen X. Production, modification and bio-applications of magnetic nanoparticles gestated by magnetotactic bacteria. *Nano Research*. 2009 Apr;2(4):261–78.

Yetisgin A.A., Cetinel S., Zuvin M., Kosar A., Kutlu O. Therapeutic nanoparticles and their targeted delivery applications. *Molecules*. 2020 Jan;25(9):2193.

Zheng D., Hu C., Gan T., Dang X., Hu S. Preparation and application of a novel vanillin sensor based on biosynthesis of Au–Ag alloy nanoparticles. *Sensors and Actuators B: Chemical*. 2010 Jun 30;148(1):247–52.

7 Advantages of Advanced Carbon-Based Nanomaterials in Medical and Pharmaceuticals

M. K. Verma and Rashmi Chowdhary

CONTENTS

7.1 CARBON-BASED NANOMATERIALS: AN OVERVIEW

Carbon is one of the most indispensable and unique elements on planet earth. Carbon is an essential element in biology and remains associated with various allotropes (Zhang and Yin 2012). It remains a critical element for synthesizing various nanomaterials, primarily carbon nanotubes, fullerenes, activated carbon, and graphite in conjugated form. For biomedical applications, carbon nanotubes and fullerenes have shown tremendous scope over the past decade. These conjugated forms of carbon resulting from diverse nano-materials differ in physicochemical properties and offer a wide range of medical and biomedical applications (Deng et al. 2016). The selective carbon-based nanomaterials, including carbon nanotubes and fullerenes, often offer the development of diagnostics and delivery systems for therapeutic cargo. Those nanostructures in the nano-scale range (10^{-9} m) serve as an ideal vehicle for drug delivery to the cellular and sub-cellular environment.

In addition, due to the unique characteristics of carbon, catenation provides an opportunity to develop a nanomaterial of different shapes and sizes. These nanomaterials are involved with part of the development of diagnostic tools and devices (Kruger 2010). Carbon-based nanomaterials allow imprinting alone or in conjugation

DOI: 10.1201/9781003110781-7

with other materials that offer an ideal platform for diagnostics (Bianco et al. 2005). Additionally, carbon-based nanomaterials are being used to develop micromachines, i.e., nanomotors and nanorobots, for various medical and biomedical applications (Worsley et al. 2009).

In recent times, carbon nanotubes (CNTs), both single-walled carbon nanotubes (SWCNTs) and multi-walled carbon nanotubes (MWCNTs), have become integral parts of medical and biomedical applications (Saito et al. 1998). It has been reported that these carbon-based nanomaterials possess additional physicochemical properties, such as shape, size, and biocompatibility, appropriate to diagnostic and therapeutic applications. There are several approaches to producing these nanocomposites for the desired application, primarily *via* synthesis. Additional properties can be introduced *via* functionalization and conjugation with other materials and groups. The vital physicochemical properties of CNTs associated with applications depend on the shape, size, and surface area that can be optimized *via* conjugation, either covalent or non-covalent modifications (Ajayan et al. 2001). In both cases, there is a greater risk of toxicity, acute and/or chronic. CNT toxicity has been affected by many physicochemical properties (e.g. size, type of functionalization), concentration, duration of exposure, mode of exposure, and even the solvents/medium used to dissolve/disperse CNTs for their application (Bianco et al. 2005). The green synthesis of CNTs and their purification, along with optimization of functional groups, are a few approaches to reduce or overcome CNT-based toxicity in the living system.

7.2 APPLICATION OF CARBON-BASED NANOMATERIALS

The advantage of carbon-based nanomaterials over other nano-designs can be assessed *via* its diverse medical and biomedical applications. The physicochemical properties of carbon-based nanomaterials, including thermal, mechanical, electrical, optical, and structural properties, facilitate a wide range of medical and biomedical applications (Adeli et al. 2013, Babele et al. 2021). In general, carbon-based nanomaterials are being used for diagnostic development and cargos for drug delivery. CNTs are microscopic, rolled-up structures of single sheets of SWCNTs and multiple sheets of MWCNTs of graphene and individual carbon structures. The CNTs, including SWCNTs and MWCNTs, remain key nanostructures in diagnostics/biosensor designs and idealized therapeutic carriers (Bhattacharya et al. 2016, Sobha et al. 2018).

7.3 CARBON-BASED NANOMATERIALS IN DIAGNOSTICS

In developing diagnostics and biosensors, the inherent physicochemical properties of carbon-based nanomaterials offer advantages over other materials. In CNTs, native physicochemical properties, including high conductivity, high chemical stability and sensitivity, and fast electron transfer rate, offer an advantage in terms of diagnostics and biosensor design (Amenta and Aschberger 2015). Furthermore, CNTs possess a vast surface area, and a greater degree of immobilization enabling these carbon-based nanomaterials to show enhanced recognition, while the signal transduction process is essential for a diagnostic or biosensor to offer useful and robust outcomes, so that CNTs, including both SWCNTs and MWCNTs, are ideal candidates for development of diagnostics (Chen et al. 2018). The CNT-based biosensors can be categorized into electrochemical, electronic, or optical biosensors, where target recognition and signal transduction are essential criteria for classification and the effectiveness of a biosensor. In carbon-based biomaterial development, a biosensor has an advantage due to native properties fulfilling all primary essential criteria (Scott et al. 2001). CNTs contain several criteria as the most promising materials for electrochemical and electronic biosensor design among different carbon-based nanomaterials (Wang et al. 2016, Mehra et al. 2014). For example, several CNT-based glucose biosensors have been designed in conjugation with the enzyme glucose oxidase, offering efficacy and robustness.

Carbon-based nanomaterials provide ease in probing, and, as a result, various biosensors for different biological parameters have been designed. In a glucose biosensor opting for an electrochemical platform, it was MWCNTs, as a key conjugate with glucose oxidase, that offers such a large surface area, increasing its efficacy (Jain et al. 2011). Similarly, SWCNT- based biosensors for human plasma protein determination have been developed. The size and diameter of CNTs serve as outstanding features in protein biosensor design for quantification of plasma/serum proteins, such as fibrinogen. The fluorescent probes and chemoselective bio-imaging systems utilize CNTs, mainly SWCNTs, for biosensor design. In conjugation with fluorescent probes, the SWCNTs offer an ideal biosensor for nucleic acid estimation in biological samples (Andon et al. 2013). The significant physicochemical properties of carbon-based nanomaterials, such as fluorescence emission in the near infra-red region and excellent photostability, are often used in designing biosensors for serum lipid profiles using the optical platform (Dinesh et al. 2016). The efficacy and robustness of CNTs, including both SWCNT- and MWCNT- based biosensors, depend largely on the absorption of light, electron flow, and thermal properties (Bilalis et al. 2014). These properties can be further optimized using functionalization, purification, and conjugation.

Graphene as graphene oxide, another carbon-based nanomaterial, has attracted attention in diagnostics/biosensor design. It does offer several additional advantages due to specific physical properties, such as ease of probing and rapid transduction response, i.e., fluorescence, Raman scattering, and electrochemical reaction (Postema and Gilja 2011). There is growing evidence of fluorescence-tagged graphene oxide being used in sensors for nucleic acids (DNA and different classes of RNAs). Functionalized graphene oxide, with specific groups and probes, also offer selective sensing of DNA, i.e., single-stranded *vs* double-stranded) and various forms of RNA (Zanganeh et al. 2013).

Furthermore, tremendous work is underway to develop sensors to detect nucleic acids and their building blocks, i.e., purine and pyrimidine bases. The oligonucleotides differ in their absorption spectra. Hence, the design of new-generation graphene oxide-based biosensors for microbial nucleic acid-sensing provides an easy and selective infection diagnosis (Diao et al. 2012). Here, graphene-Si-nanowires provide extended glucose, nucleic acid, and lipid profiles, and microbial detection with precision (Jena et al. 2016).

7.4 CARBON-BASED NANOMATERIALS IN MEDICINE

In recent times, nanotechnology has become one of the most promising areas, offering many medical applications. Over time, several classes of nanomaterials have been designed and evaluated for their use in medicine. The carbon-based nanomaterials, including nanoparticles, carbon nanotubes, graphene, and many more, are critical carriers for drug delivery to cellular and sub-cellular levels (Brown and Semelka 2011). Overwhelming research evidence demonstrates that CNTs serve as the most versatile cargoes for drug delivery for cancer management. However, carbon-based nanomaterial in medicine is limited to drug delivery for cancer management and several other diseases and pathological conditions (Raffa et al. 2008).

On the contrary, nanotechnology, which include carbon-based nanomaterials, is extensively used to diagnose cancer and to deliver various anticancer agents. The superiority of carbon-based nanomaterials over other nanomaterials as drug delivery agents in terms of the physical properties of nanomaterials such as CNTs (SWCNTs and MWCNTs) and the capacity to functionalize/modify materials readily (Reddy et al. 2006). The fundamental physical properties, including thermal and electrical conductivity, aspect ratio, tensile strength, and thermal expansion coefficient, distinguish CNTs from other nanomaterials. Furthermore, shape/size, solubility, an affinity for a particular cell/tissue (native and/or functionalized), biodegradation and biocompatibility are additional physical properties of value for drug delivery systems (Yang et al. 2008).

The use of carbon-based nanomaterials, such as CNTs, as drug delivery cargo is one of the most promising and well-studied nanotechnology areas. Nanomaterials, including carbon-based nanomaterials, were designed and evaluated for anticancer drug delivery in the past few decades. These nanomaterials offer delivery of a wide range of therapeutic agents, including anticancer drugs, genes, nucleic acids, and proteins to a targeted tissue or cell type as well (Yang et al. 2009). Here, the targeted delivery remains a function of the nanomaterial shape/size and affinity for the tissue/cell. Delivery can be further improved *via* optimization of the nanomaterial, including functionalization and conjugation. Compared with liposomes/immunoliposomes, carbon-based nanomaterials, e.g., CNTs, are significant in carrying a large cargo of therapeutics and can be monitored readily in real-time (Liu et al. 2003). The native physical properties of carbon nanomaterials are often used for selective medical and biomedical applications. For example, graphene has shown a reduced cytotoxicity level, improved hydrophilicity, stable intrinsic fluorescence property, and surface functional groups that serve as ideal carriers for pharmaceuticals and radiopharmaceuticals for cancer diagnostic and treatment (Yang et al. 2011). Furthermore, carbon-based nanomaterials also offer photothermal- and ultrasonic-based drug release to a targeted tissue and/or cell, offering real-time monitoring of drug delivery.

Cancer is a multifactorial disorder that requires a selective and effective intervention with therapeutics for its management. In the past few decades, progressive and new-generation approaches have been developed to diagnose and treat cancer (Sahoo et al. 2011). Compared with conventional cancer diagnosis methods and treatment, carbon nanomaterial-based drug delivery showed tremendous scope for cancer management. The high degree of functionalization of carbon-based nanomaterials, such as CNTs, offers a wide range of anticancer therapeutics (Dhar et al. 2008). CNTs and other carbon-based nanomaterials provide an opportunity for both covalent and non-covalent functionalization. It has been shown that non-carbon-based nanomaterial often fails to functionalize and conjugate as well. Several human organs are well protected and have other membranes, such as neural tissue protected by the blood-brain barrier (BBB) (Hampel et al. 2008). Conventional drug delivery methods directed towards such tissues often fail, and the excess dose may lead to toxic effects of the drugs. Carbon-based nanomaterials, mainly CNTs, are ideal for effective and selective drug delivery across the blood-brain barrier (Klibanoy et al. 1990). Overwhelming research evidence suggested that CNTs, both SWCNTs, and MWCNTs, remain key drug delivery cargos for chemotherapies and photothermal therapies for cancer management (Allen et al. 1992).

Drug release from cargoes, such as CNTs, liposomes, and immunoliposomes, must be regulated to cause effective therapy. CNTs have shown an advantage over other nano-composites in the ordered release of loaded therapeutics. A disordered drug release from given cargoes may end with a small volume of distribution (Vd), leading to the accumulation of the drug in the tissue and subsequent toxicity (Li et al. 1996). The large surface area of any nanomaterial/nanodesign is a crucial prerequisite for ideal drug delivery vehicles. The shape and size of nanomaterials are vital for exhibiting significant pharmacokinetics, i.e., absorption, distribution, metabolism, and excretion (ADME). Increasing research evidence suggests that carbon-based nanomaterials are far superior to non-carbon-based nanomaterials in acquiring ADME properties and, more precisely, in achieving release/distribution of the loaded drug (Liu et al. 2008). The large surface area of graphene and its capacity to functionalize offers several benefits, including a wide range of drugs, hydrophilic and hydrophobic, to be delivered, and optimal ADME properties. The anticancer agents differ in their chemical structure, and it remains a major challenge to develop a common and effective platform to accommodate a large range of anticancer agents. Growing research demonstrates that graphene is an ideal platform for indomethacin and doxorubicin, which are two hydrophobic anticancer drugs (Lay et al. 2010).

7.5 OTHER MEDICAL AND BIOMEDICAL APPLICATIONS

The application of carbon-based nanomaterials is not limited to developing diagnostics/biosensor, as a vehicle for drug delivery, or to other medical and biomedical areas (Silver and Landis 2011). The role of carbon-based nanomaterials in tissue engineering has been well studied and reported. Such nanostructures are ideal scaffolds for tissue generation *in vitro*. The benefit of using a carbon-based nanomaterial as scaffolds is that it allows diverse sizes and shapes for tissue and organ generation. Additionally, these scaffolds are often inert and can be used repeatedly (Mcnally et al. 2015). Research demonstrated that carbon-based nanomaterials, including graphene, graphene oxide, and CNTs, provide an ideal scaffold for *in-vitro* generation of cardiac, neural, skin, or bone tissue (Zhao et al. 2005). Most of the studies associated with CNT-based tissue regeneration suggest highly biocompatible materials and accelerated osteogenesis. Compared with SWCNTs, MWCNTs are more biocompatible with bone tissue, exhibit greater absorption and encourage cellular development and differentiation (Lobo et al. 2011). For a carbon-based nanomaterial to serve as a scaffold for tissue regeneration requires rigidity to bear external force, biodegradability, and absorption, and the ability to promote the adhesion and proliferation of cells and the ability to be penetrated by blood vessels and body fluids (Sciortino et al. 2017). Studies have shown that functionalization and conjugation of MWCNTs causes additional properties to be acquired that promote cellular growth, development, and differentiation.

The use of carbon-based nanomaterials in dentistry has gained tremendous scope in the last few years, on multiple forms as conjugated and functionalizedas needed (Stoller et al. 2008). The use of carbon nanomaterials in dentistry, more precisely nano-dentistry, remains associated with chemical vapour deposition (CVD), monolithic processing, and wet and plasma etching (Hu et al. 2010). Carbon-based nanomaterials offer additional protection to teeth and gum compared to conventional composite materials. The research findings have demonstrated that carbon nanomaterial-based dental implants, cavity filling material, and gum treatments offer extended protection to tissue along with enhanced vascular flow (Carpio et al. 2012). Applications of the nanomaterial in dentistry, including carbon-based nanomaterials, includes hypersensitivity care, local anaesthesia for surgical intervention, orthodontic treatment, and nano-impressions (Misch 2014).

Furthermore, nanomaterial-based dental products are typical for oral hygiene and dental health as well. These materials are helpful in the early diagnosis of gum diseases, including chronic gum inflammation and cancer (Kulshrestha et al. 2014). The carbon-based nanomaterials are also part of dental formulations (toothpaste and mouthwashes), such as subocclusal nanorobotic dentifrices, i.e. dentifrobots as nanorobotic dentifrice (Bartolo et al. 2012). Such oral formulations containing carbon-based nanomaterials for dental use have shown an increasing scope to ensure oral and dental health.

7.6 TOXICITY OF CARBON-BASED NANOMATERIALS

Carbon-based nanomaterials, including CNTs, have promising applications in almost all walks of life, with data having been obtained about the use of CNTs for cancer treatment (Ferreira et al. 2014). Various research groups have employed a range of techniques *in vivo* and *in vitro* to determine the factors that lead to CNT-mediated cytotoxicity. Different methods are used to assess this toxicity, such as the Alamar blue, the tetrazolium-based MTT, and the Trypan blue assays (Kafa et al. 2015). CNTs can significantly disrupt the emission of specific particles, which are detected by different methods. In contrast, in *in-vivo* studies, the measurement of toxicity and pathology caused by CNTs varies from cell to cell and from organism to organism (Kagan et al. 2014). Despite the various drawbacks of *in vitro* studies, a detailed CNT toxicity analysis is fundamental to providing new insights into this emerging area. However, the use of these nanomaterials for various biomedical services also exposes human beings to their harmful impact. Such risks are increasing day by day due to the increasing use of these nanomaterials and devices to improve the quality of human life (Kang et al. 2010).

Hence, it is of the utmost importance to explore the toxicity issues of these new nanomaterials. A number of studies have been conducted on the toxicology, clearance, or biodistribution of various nanomaterials, such as metal and metal oxide nanoparticles, carbon nanoparticles and nanotubes, quantum dots, and magnetic nanoparticles and any risks they pose to humans and the environment. Cells exposed to higher concentrations, i.e., >400 μg mL^{-1} of CNTs, lead to programmed cell death, whereas lower concentrations of CNTs were not nearly as toxic (Kang et al. 2007, Kim et al. 2014, Lan et al. 2014). The potential hazards of CNTs on human health and the environment are of great concern as the advances in the synthesis and applications of CNTs in every walk of life continues. Much research is going on to analyze CNT toxicity at the cellular and molecular levels, and even on whole animals, but the findings of these studies are often contradictory (Lan and Yang 2012).

Despite increasing commercial use of CNTs, the associated toxicity remains a significant issue. The lack of a collective database for CNTs and the varying chemical nature of CNTs inhibits the development of a standard analytical method to evaluate associated toxicities. CNT-associated toxicity has been a primary concern from the beginning for the safe use of nanotechnology in medicine, agriculture, environments, and other industries (Lee et al. 2014). It is essential to understand the source of toxicity in CNTs before developing an analytical approach and method. Several components can contribute to toxicity associated with CNTs, including inorganic residues used as a catalyst, the nature and physicochemical properties of CNTs, toxicity due to fabrication and functionalization, and impurities in CNTs and toxicity due to chemicals introduced for enhancement of solubility (Levine and Klionsky 2004). Considering all these points, the evaluation of CNT-associated toxicity depends largely on *in-vivo* and *in-vitro* analyses (Verma et al. 2013). It is essential to note that the evaluation of CNT toxicity ultimately depends on the tissue, the animal model, and the CNTs (particularly their nature and physicochemical properties) (Lindberg et al. 2009). A number of cutting-edge biological and biochemical methods, such as transcriptomics (patterns of gene expression) proteomics (protein patterns), and metabolomics (assessed by GC, HPLC, GC-MS, or LC-MS) are in routine use in the evaluation of CNT toxicity (Ma et al. 2012).

7.7 FUTURE DIRECTIONS

An extensive research effort has explored the medical and biomedical applications of carbon-based nanomaterials and nanodevices over the past couple of decades. However, despite the increasing use of these nanomaterials and nanodesigns, toxicity remains a key concern. Future research should be directed to minimize such toxic effects and their impact on living tissues/cells. There is much emphasis on understanding the nature and mechanism of toxicity caused by carbon-based nanomaterials. The application of carbon nanomaterials is primarily centred on the delivery of antineoplastic agents, and hence more attention should be focused on other class of drugs/

therapeutics. Therefore, future research is necessary to optimize carbon-based nanomaterials and nanodesigns for safe and effective drug delivery. Drug delivery across the blood-brain barrier (BBB) remains a significant challenge. Studies have shown that CNTs can cross blood-brain barriers and possibly treat neurological disorders, such as cerebral ischaemia, Parkinson's disease, Alzheimer's disease, multiple sclerosis, etc.

Furthermore, several studies have raised concerns that CNTs can be an ideal fabricated material for diagnostic and therapeutic applications. CNTs are challenging to solubilize (in biological fluids), are non-biodegradable and biocompatible, and pose immunogenicity issues. Apart from the biological applications, the synthesis and fabrication level remains a significant challenge for CNT production. The higher percentage of self-assembling and random distribution of CNTs results in the lowering of desirable physical properties.

7.8 SUMMARY

Carbon nanotubes (CNTs) possess many distinct properties, including good electronic properties, remarkable penetrating capability towards cell membranes, high drug loading and pH-dependent therapeutic unloading capacities, thermal properties, large surface area, and ease of modification with molecules, rendering them suitable candidates to deliver drugs to cancer and the brain. The CNTs, along with graphene and graphene oxides, are primarily used for biosensor and drug delivery vehicles. These carbon-based nanomaterials and designs provide an ideal platform for biosensor development with high precision and accuracy. These materials and designs are already revolutionizing diagnostics/biosensors and drug delivery systems, but the increasing medical and biomedical applications of carbon-based nanomaterials and nanodesigns are also associated with toxicity to living tissue. Ongoing research work must focus on strategies by which to minimize such toxicity for an extended application. In addition, challenges remain, such as drug delivery to the brain *via* the blood-brain barrier (BBB) and how to measure and minimize CNT-associated toxicity.

7.9 REFERENCES

Adeli, M., Soleyman, R., Beiranvanda, Z., Madani, F. 2013. Carbon nanotubes in cancer therapy: a more precise look at the role of carbon nanotube-polymer interactions. *Chem. Soc. Rev.* 42 5231–5256. 10.1039/c3cs35431h

Ajayan, P. M., Zhou, O. Z. 2001. Applications of carbon nanotubes. In: Dresselhaus, M. S., Dresselhaus, G., Avouris, P., editors. Carbon Nanotubes: Synthesis, Structure, Properties, and Applications. Springer-Verlag; New York: pp. 391–425.

Allen, T. M., Mehra, T., Hansen, C., Chin, Y. C. 1992. Stealth liposomes: an improved sustained release system for 1-β-D-arabinofuranosylcytosine. *Cancer Research.* 52(9):2431–2439.

Amenta, V., Aschberger, K. 2015. Carbon nanotubes: potential medical applications and safety concerns. *Adv. Rev.*7 371–386. 10.1002/wnan.1317

Andón, F. T., Kapralov, A. A., Yanamala, N., Feng, W., Baygan, A., Chambers, B. J., Hultenby, K., Ye, F., Toprak, M. S., Brandner, B. D., et al. 2013. Biodegradation of Single-Walled Carbon Nanotubes by Eosinophil Peroxidase. Small. 9:2721–2729. doi: 10.1002/smll.201202508

Babele, P. K., Verma, M. K. and Bhatiya, R. 2021. Carbon Nanotubes: A review on risks assessment, mechanism of toxicity and future directives to prevent health implication, Bio-Cell 42 (2) 267–279

Bartolo, P., Kruth, J. P., Silva, J. 2012. Biomedical production of implants by additive electro-chemical and physical processes. *CIRP Annals - Manufacturing Technology 61*, 635–655.

Bhattacharya, K., Mukherjee, S. P., Gallud, A., Burkert, S. C., Bistarelli, S., Bellucci, S., et al. 2016. Biological interactions of carbon-based nanomaterials: from coronation to degradation. *Nanomedicine.* 12 333–351. 10.1016/j.nano.2015.11.011

Bianco, A., Kostarelos, K., Partidos, C. D., Prato, M. 2005. Biomedical applications of functionalized carbon nanotubes. *Chem Comm.* 571–577.

Bilalis, P., Katsigiannopoulos, D., Avgeropoulos, A., Sakellariou, G. 2014. Non-Covalent functionalization of carbon nanotubes with polymers. Rsc Adv. 4:2911–2934. doi: 10.1039/C3RA44906H

Brown, M. A., Semelka, R. C. 2011. MRI: Basic Principles and Applications. John Wiley & Sons; London, UK.

Carpio, I. E. M., Santos, C. M., Wei, X., Rodrigues, D. F. 2012. Toxicity of a polymer–graphene oxide composite against bacterial planktonic cells, biofilms, and mammalian cells. Nanoscale. 4:4746–4756.

Chen, F., Gao, W., Qiu, X., Zhang, H., Liu, L., Luo, Y. 2018. Graphene quantum dots in biomedical applications: recent advances and future challenges. *Front. Lab. Med.* 1; 192–199. 10.1016/j.flm.2017.12.006

Deng, J., You, Y., Sahajwalla, V., Joshi, R. K. 2016. Transforming waste into carbon-based nanomaterials. *Carbon 96*, 105–115.

Dhar, S., Liu, Z., Thomale, J., Dai, H., Lippard, S. J. 2008. Targeted single-wall carbon nanotube-mediated Pt(IV) prodrug delivery using folate as a homing device. *Journal of the American Chemical Society.* 130(34):11467–11476.

Diao, S., Hong, G., Robinson, J. T., Jiao, L., Antaris, A. L., Wu, J. Z., Choi, C. L., Dai, H.2012. Chirality enriched (12, 1) and (11, 3) single-walled carbon nanotubes for biological imaging. J. Am. Chem. Soc. 134:16971–16974. doi: 10.1021/ja307966u

Dinesh, B., Bianco, A., Ménard-Moyon, C. 2016. Designing multimodal carbon nanotubes by covalent multi-functionalization. Nanoscale. 8:18596–18611. doi: 10.1039/C6NR06728J.

Ferreira, J., Pires, P. T., Almeida, C., Jerónimo, S., and Melo, P. 2014. Avaliação da Eficácia do nanoXIM Care Paste na Oclusão dos Túbulos Dentinários/Evaluation of the Efficacy of nanoXIM CarePaste in Dentinal Tubule Occlusion. In International Poster Journal of Dentistry and Oral Medicine - XXIII Congresso OMD; Matosinhos, Portugal.

Hampel, S., Kunze, D., Haase, D., et al. 2008. Carbon nanotubes filled with a chemotherapeutic agent: a nanocarrier mediates inhibition of tumor cell growth. *Nanomedicine.* 3(2):175–182.

Hu, W., Peng, C., Luo, W. 2010. Graphene-based antibacterial paper. ACS Nano.4:4317–4323.

Jain, S., Thakare, V. S., Das, M., Godugu, C., Jain, A. K., Mathur, R., Chuttani, K., Mishra, A. K. 2011. Toxicity of multi-walled carbon nanotubes with end defects critically depends on their functionalization density. Chem. Res. Toxicol. 24:2028–2039. doi: 10.1021/tx2003728

Jena, P. V., Shamay, Y., Shah, J., Roxbury, D., Paknejad, N., Heller, D. A. 2016. Photoluminescent carbon nanotubes interrogate the permeability of multicellular tumor spheroids. Carbon. 97:99–109. doi: 10.1016/j.carbon.2015.08.024

Kafa, H., Wang, J. T.-W., Rubio, N., Venner, K., Anderson, G., Pach, E., Ballesteros, B., Preston, J. E., Abbott, N. J., and Al-Jamal, K. T. 2015. The interaction of carbon nanotubes with an in vitro blood-brain barrier model and mouse brain in vivo. *Biomaterials. 53*, 437–452.

Kagan, V. E., Kapralov, A. A., St. Croix, C. M., Watkins, S. C., Kisin, E. R., Kotchey, G. P., Balasubramanian, K., Vlasova, I. I., Yu, J., and Kim, K. Lung macrophages "digest" carbon nanotubes using a superoxide/peroxynitrite oxidative pathway. *ACS Nano*; *8*, 5610–5621.

Kang, B., Chang, S., Dai, Y., Yu, D., and Chen, D. 2010. Cell response to carbon nanotubes: Size-dependent intracellular uptake mechanism and subcellular fate. *Small 6*, 2362–2366.

Kang, S., Pinault, M., Pfefferle, L. D., and Elimelech, M. 2007. Single-walled carbon nanotubes exhibit strong antimicrobial activity. *Langmuir 23*, 8670–8673.

Kim, K. H., Yeon, S.-m., Kim, H. G., Lee, H., Kim, S. K., Han, S. H., Min, K.-J., Byun, Y., Lee, E. H., and Lee, K. S. 2014. Single-walled carbon nanotubes induce cell death and transcription of TNF-α in macrophages without affecting nitric oxide production. Inflammation *37*, 44–54.

Klibanov, A. L., Maruyama, K., Torchilin, V. P., Huang, L. 1990. Amphipathic polyethylene glycols effectively prolong the circulation time of liposomes. *FEBS Letters*. 268(1):235–237.

Kruger, A. 2010. *Carbon Materials and Nanotechnology*. Wiley-VCH; Weinheim.

Kulshrestha, S., Khan, S., Meena, R., Singh, B. R., Khan, A. U. 2014. A graphene/zinc oxide nanocomposite film protects dental implant surfaces against cariogenic Streptococcus mutans. Biofouling. 30:1281–129

Lan, J., Gou, N., Gao, C., He, M., and Gu, A. Z. 2014. Comparative and mechanistic genotoxicity assessment of nanomaterials via a quantitative toxicogenomics approach across multiple species. *Environmental Science & Technology 48*, 12937–12945.

Lan, Z., and Yang, W. X. 2012. Nanoparticles and spermatogenesis: how do nanoparticles affect spermatogenesis and penetrate the blood–testis barrier. *Nanomedicine 7*, 579–596.

Lay, C. L., Liu, H. Q., Tan, H. R., Liu, Y. 2010. Delivery of paclitaxel by physically loading onto poly(ethylene glycol) (PEG)-graftcarbon nanotubes for potent cancer therapeutics. *Nanotechnology*. 21(6) Articles ID 065101.

Lee, Y. H., Cheng, F. Y., Chiu, H. W., Tsai, J. C., Fang, C. Y., Chen, C. W., and Wang, Y. J. 2014. Cytotoxicity, oxidative stress, apoptosis and the autophagic effects of silver nanoparticles in mouse embryonic fibroblasts. *Biomaterials 35*, 4706–4715.

Levine, B., and Klionsky, D. J. 2004. Development by self-digestion: molecular mechanisms and biological functions of autophagy. *Developmental Cell 6*, 463–477.

Li, C., Yu, D., Inoue, T., et al. 1996. Synthesis and evaluation of water-soluble polyethylene glycol-paclitaxel conjugate as a paclitaxel prodrug. *Anticancer Drugs*. 7(6):642–648.

Lindberg, H. K., Falck, G. C.-M., Suhonen, S., Vippola, M., Vanhala, E., Catalán, J., Savolainen, K., and Norppa, H. 2009. Genotoxicity of nanomaterials: DNA damage and micronuclei induced by carbon nanotubes and graphite nanofibres in human bronchial epithelial cells in vitro. *Toxicology Letters 186*, 166–173.

Liu, Y., Ng, K. Y., Lillehei, K. O. 2003. Cell-mediated immunotherapy: a new approach to the treatment of malignant glioma. *Cancer Control*. 10(2):138–147.

Liu, Z., Chen, K., Davis, C., et al. 2008. Drug delivery with carbon nanotubes for in vivo cancer treatment. *Cancer Research*. 68(16):6652–6660.

Lobo, A. O., Marciano, F. R., Regiani, I., Ramos, S. C., Matsushima, J. T., Corat, E. J. 2011. Proposed model for growth preference of plate-like nanohydroxyapatite crystals on superhydrophilic vertically aligned carbon nanotubes by electrodeposition. Theor. Chem. Acc. 130:1071–1082. doi: 10.1007/s00214-011-0993-x. 58.

Ma, X., Zhang, L.-H., Wang, L.-R., Xue, X., Sun, J.-H., Wu, Y., Zou, G., Wu, X., Wang, P. C., and Wamer, W. G. 2012. Single-walled carbon nanotubes alter cytochrome c electron transfer and modulate mitochondrial function. *ACS Nano 6*, 10486–10496.

Mcnally, E. A., Schwarcz, H. P., Botton, G. A., Arsenault, A. L. 2015. A model for the ultrastructure of bone based on electron microscopy of ion-milled sections. PLoS ONE. 46:44–50. doi: 10.1371/journal.pone.0029258.

Mehra, N. K., Mishra, V., Jain, N. K. 2014. A review of ligand tethered surface engineered carbon nanotubes. Biomaterials. 35:1267–1283. doi: 10.1016/j.biomaterials.2013.10.032.

Misch, C. E. 2014. Elsevier Health Sciences. Dental Implant Prosthetics-E-Book.

Postema, M., Gilja, O. H. 2011. Contrast-Enhanced and targeted ultrasound. World J. Gastroenterol. WJG. 17:28. doi: 10.3748/wjg.v17.i1.28.

Raffa, V., Ciofani, G., Nitodas, S., et al.2008. Can the properties of carbon nanotubes influence their internalization by living cells? *Carbon*. 46(12):1600–1610.

Reddy, S. T., Rehor, A., Schmoekel, H. G., Hubbell, J. A., Swartz, M. A. 2006. In vivo targeting of dendritic cells in lymph nodes with poly(propylene sulfide) nanoparticles. *Journal of Controlled Release*. 112(1):26–34.

Sahoo, N. G., Bao, H., Pan, Y., et al. 2011. Functionalized carbon nanomaterials as nanocarriers for loading and delivery of a poorly water-soluble anticancer drug: a comparative study. *Chemical Communications*. 47(18):5235–5237.

Saito, R., Dresselhaus, G., Dresselhaus, M. S. 1998. Physical Properties of Carbon Nanotubes. Imperial College Press; London.

Sciortino, N., Fedeli, S., Paoli, P., Brandi, A., Chiarugi, P., Severi, M., Cicchi, S. 2017. Multiwalled carbon nanotubes for drug delivery: Efficiency related to length and incubation time. Int. J. Pharm. 521:69–72. doi: 10.1016/j.ijpharm.2017.02.023.

Scott, C. D., Arepalli, S., Nikolaev, P., Smalley, R. E. 2001. Growth mechanisms for single-wall carbon nanotubes in a laser-ablation process. Appl. Phys. A. 72:573–580. doi: 10.1007/s003390100761.

Silver, F. H., Landis, W. J. 2011. Deposition of apatite in mineralizing vertebrate extracellular matrices: A model of possible nucleation sites on type I collagen. Connect. Tissue Res. 52:242–254. doi: 10.3109/03008207.2010.551567.

Sobha, K., Pradeep, D., Kumari, A. R., Verma, M. K. and Surendranath, K. 2018. Evaluation of the biological activity of the silver nanoparticles synthesized with the aqueous leaf extract of *Rumex acetosa*, Scientific Reports 7, 11566, DOI:10.1038/s41598-017-11853-2

Stoller, M. D., Park, S., Zhu, Y., An, J., Ruoff, R. S. 2008. Graphene-based ultracapacitors. Nano Lett.8:3498–3502.

Verma, M. K., Verma, Y. K., Kota, S., and Dey, S. K. 2013. Current prospects of nano-designs in gene delivery-aiming new high for efficient and targeted gene therapy. *International Journal of Biopharmaceutics 4* (3), 146–165.

Wang, Z., Yu, J., Gui, R., Jin, H., Xia, Y. 2016. Carbon nanomaterials-based electrochemical aptasensors. Biosens. Bioelectron. 79:136–149. doi: 10.1016/j.bios.2015.11.093

Worsley, M. A., Kucheyev, S. O., Kuntz, J. D., Hamza, A. V., Satcher, J. J. H., Baumann, T. F. 2009. Stiff and Electrically Conductive Composites of Carbon Nanotube Aerogels and Polymers. J Mater Chem. 19:3370–3372.

Yang, F., Fu, D. L., Long, J., Ni, Q. X. 2008. Magnetic lymphatic targeting drug delivery system using carbon nanotubes. *Medical Hypotheses*. 70(4):765–767.

Yang, F., Hu, J., Yang. D., et al. 2009. Pilot study of targeting magnetic carbon nanotubes to lymph nodes. *Nanomedicine*. 4(3):317–330.

Yang, F., Jin, C., Yang, D., et al. 2011. Magnetic functionalized carbon nanotubes as drug vehicles for cancer lymph node metastasis treatment. *European Journal of Cancer*. 47(12):1873–1882.

Zanganeh, S., Li, H., Kumavor, P. D., Alqasemi, U., Aguirre, A., Mohammad, I., Stanford, C., Smith, M. B., Zhu, Q. 2013. Photoacoustic imaging enhanced by indocyanine green-conjugated single-wall carbon nanotubes. J. Biomed. Opt. 18:096006. doi: 10.1117/1.JBO.18.9.096006.

Zhang, Y., Yin, Q. Z. 2012. Carbon and other light element contents in the Earth's core based on first-principles molecular dynamics. *Proceedings of the National Academy of Sciences of the United States of America 109*, 19579–19583.

Zhao, B., Hui, H., Mandal, S. K., Haddon, R. C. 2005. A Bone Mimic Based on the Self-Assembly of Hydroxyapatite on Chemically Functionalized Single-Walled Carbon Nanotubes. Chem. Mater. 17:3235–3241. doi: 10.1021/cm0500399.

8 Carbon Nanotubes/ Graphene-Based Chemiresistive Biosensors

Suresh Bandi and Ajeet K. Srivastav

CONTENTS

8.1 INTRODUCTION

Biosensors are tiny devices that play an essential role in the safety, disease detection, and contaminant detection of various fields viz., food processing, beverages, groundwater, environment, agriculture, bioterrorism, and health sectors. Indirectly, all the abovementioned fields significantly impact the health sectors with their hidden negative effects on human society. To understand the seriousness, let's elaborate on them. The food/beverage processing includes several steps, from preparation to preservation. The unbalanced compositions of fats, sugars, vitamins, pesticides, food contaminants, food-borne pathogens/microorganisms, toxins, etc., can cause a severe adverse effect on human health.

Water or soil contamination is another threat caused by industrial wastes, agricultural pesticides, military explosives, etc. For example, 2,4,6-trinitrotoluene (TNT) is a compound used in industrial applications and military explosives. Water and soil contamination by this compound has been found in the regions surrounding chemical plants and where explosion activities had occurred. It is a carcinogenic compound that causes anaemia, liver infection, skin diseases, and weakening of the

DOI: 10.1201/9781003110781-8

immune system. Several conventional techniques are followed by government organizations for TNT contaminant monitoring or quality check of all the mentioned fields. However, there is a requirement for faster, reliable, and inexpensive portable devices for such analysis. In the healthcare sector, these sensors play a valuable role in the early detection of a number of diseases, especially for the detection of emerging infectious diseases (EID) (Pejcic et al. 2006). Biosensors are inexpensive devices which a normal human being can also afford to use. With this advance, people have the ability at their fingertips to instantaneously assess the quality of what they are eating or drinking, or to monitor several biological fluids for the initial identification of a number of diseases.

However, as the sensitivity is derived from the sensor's resistance, the sensing material must be capable of responding to minute resistance/conductance changes during the biosensing. Hence, carbon nanotubes (CNTs) and graphene could be candidate materials for biosensors due to their excellent conductivity properties owing to their tremendous charge carrier mobility. In the current chapter, the significance of the aforementioned materials and their biosensing properties towards various biological and environmental factors are discussed.

8.2 CHEMIRESISTIVE BIOSENSORS

A chemiresistive biosensor comprises a substrate, interdigitated or microfabricated electrodes on the substrate, and a sensing material called a transducer element. Usually, the sensing material is transferred to the substrate which is printed with the electrodes using drop-casting, contact printing, spray coating, or other techniques. The sensing material bridges a conducting channel between the electrodes. The sensor is connected to a resistance measuring unit. The change in resistance/conductance in response to the analyte concentration is considered to be a measure of sensitivity. The data are shown in terms of change in the resistance *versus* concentration plot. In the chemiresistive biosensors, the role of the sensing material is transduction.

The biosensor undergoes specific modifications before the sensing studies based on the required measurements. Apart from the necessary circuitry and transducer material, there are two other vital steps to be understood. In general, the design of the biosensor is always directed to a particular, targeted cause. It starts with functionalizing transducer elements with a receptor, then ligation to the receptor of appropriate antibodies based on the targeted application. The analyte usually contains a biomarker/biomolecule based on which the selection of antibodies will be made. The antibodies selectively interact with specific biomarkers in the analyte. The antibody–biomarker interaction induces charge doping or drain in response to the bio-interaction during biosensing, where the transducer material is used to identify such a response. Hence, functionalization of the transducer material is an essential and primary step.

How is the transducer/sensing material responsible for detecting such bioreactions? The biosensor's resistance measurements are continuously monitored from the initial step of transducer element decoration, after functionalization with bioreceptors and ligated antibodies, and during the chemical/biological interactions with the biomarkers. Upon interaction of the biomarkers with the bioreceptor-functionalized sensing material, the changes that occur therein will affect the carrier concentration; the transducer element can identify these minute changes as a change in resistance/conductance. Thus, like chemiresistive gas sensing, chemiresistive biosensing also shows the difference in resistance to specific analyte interactions. Response time, sensitivity, and selectivity are the parameters that evaluate the efficiency of the sensor.

Usually, the functionalizing bioreceptors are selected based on the targeted biomarkers. It is worth mentioning that the biomarkers are always present in the specific biofluid or compound used as an analyte in the biosensing. However, the analytes will vary depending on the applications we are focused on. For example, for detecting microorganisms, the sensing material needs to be functionalized with specific antibodies tagged to a particular receptor protein by which the sensing material is already functionalized. These antibodies decide the specificity of the detection of microorganisms (Garcia-Aljaro et al. 2010). A similar strategy is followed for the detection of hazardous molecules, like TNT, as well. However, functionalization of transducer materials with appropriate antibodies for specific detection with high sensitivities is inevitable (Park et al. 2010). Any biological medium, namely saliva, serum, cellular fluid, plasma, etc., is used as an analyte to detect disease-related issues. Monitoring the level of the particular biomolecule in the aforementioned biological fluids is necessary to index the specific biological process. Hence, these biomolecules are called biomarkers. The biosensing process and data acquisition from a chemiresistive biosensor are illustrated schematically in Figure 8.1.

8.3 GRAPHENE AND CNTs: SIGNIFICANCE AND SYNTHESIS

8.3.1 Significance

Graphene and CNTs are both derivatives of graphite. In other words, graphene is a fundamental building block for the other-dimensional graphitic materials, namely fullerenes, CNTs, and graphite (Bandi and Srivastav 2020). Graphene is a 2-dimensional (2D), one-atom thick, layered, sp^2 hybridized carbon material configured in a hexagonal or honeycomb lattice (Geim and Novoselov 2007, Bandi et al. 2019), whereas a CNT is considered to be a seamless, rolled-up cylinder of a graphene monolayer, i.e., mainly termed a single-walled CNT (SWCNT) (Rao et al. 2018). Hence, both materials exhibit similar properties except for the dimensional dissimilarities (Rao et al. 2018). The high electron mobility, high surface area, excellent thermal conductivity, and exceptional Young's modulus values are unique features of these materials. Another intriguing aspect about the SWCNTs is the helical angle (χ), i.e., the angle between the tube axis and the graphene lattice (Rao et al. 2018, Tasis et al. 2006). This helical angle decides whether the CNT is semi-conducting or metallic. At the same time, the CNTs consisting of multi-walls are called multiwalled CNTs (MWCNTs). The rolling-up of multilayered graphene could form these MWCNTs. Owing to their intriguing features and properties, these materials are potential candidates for various applications, such as in energy, catalysis,

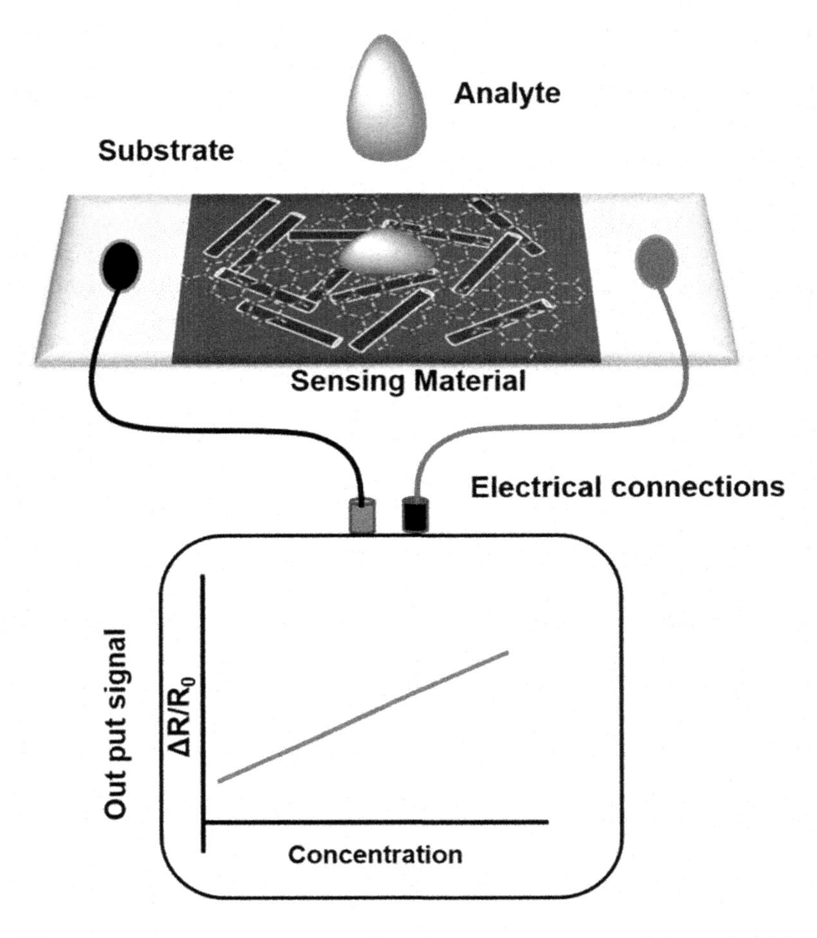

FIGURE 8.1 Schematic representation of the biosensing process and data acquisition in a chemiresistive biosensor.

environmental issues, sensors, safety, electronics, automobiles, structural engineering, agriculture, bio-medical, pharmaceutical, etc.

8.3.2 Synthesis

Graphene synthesis, in simple terms, involves separating the monolayers from graphite. The discovery of this material itself came through a micro-mechanical cleavage of graphite by Novoselov et al. (2004). Later, various advances in graphene synthesis came about because of scalability, quality, and economy. The Hummers/modified Hummers methods and exfoliation techniques are widely practised approaches for graphene synthesis (Bandi and Srivastav 2020). In the former process, graphite powder is kept in an oxidizing chemical bath atmosphere where the oxidation of graphite leads to the separation of sheets called graphene oxide (GO) (Hummers and Offeman 1958). This GO is subjected to a reduction treatment (chemical/thermal) to achieve reduced GO (rGO), i.e., graphene. The exfoliation techniques are further categorized into mechanical exfoliation, liquid-phase exfoliation, and oxidative exfoliation. The first two methods produce graphene directly. However, the scalability may not be efficient. The last approach (oxidative exfoliation) results in GO which has to undergo further reduction treatment to achieve rGO (Bandi et al. 2019). Electrochemical exfoliation is a type of oxidative exfoliation. The graphite is used as the anode, and the exfoliation happens in the electrolyte provided, with the supplied voltage bias between anode and cathode. This is one of the most practised approaches because of advantages of scalability, flexibility, and economy, combined with less chemical tediousness. In addition, other techniques, like chemical vapour deposition (CVD), template methods, arc discharge methods, epitaxial growth, and CNT unzipping, have also been adopted for the synthesis of high-quality graphene. However, the scalability is low, along with the involvement of greater costs and effort.

The synthesis of CNTs is mostly *via* gas-phase processes. The carbon arc discharge/electric arc discharge (AD), laser ablation (LA), and CVD methods are three widely practised techniques for synthesizing CNTs (Eatemadi et al. 2014). In the electric arc discharge method, graphite electrodes are used as carbon sources, where two such electrodes acting as anode and cathode are kept enclosed, between which an electric discharge is created (Mtibe et al. 2018). The current of 40–125 A and potential difference of 20–50V is supplied from the energy source. The complete process is carried out in a closed chamber with an inert atmosphere under a particular pressure. Carbon evaporation from the graphite can be achieved with metal catalysts (Co, Ni, or Fe) and high temperatures. The evaporated carbon condenses near the cathode and forms CNTs. The yield and the quality of the CNTs are two significant concerns in this approach. The laser ablation method uses laser pulses as a source of energy, and the remaining process is similar to that of AD. A piece of graphite is used as a source of carbon in this process. The laser power, catalyst, and temperature generated are the key parameters that decide the quality and yield of the CNTs. CVD is a method primarily used for CNT synthesis, which has mainly replaced the AD and LA techniques (Eatemadi et al. 2014). Hydrocarbons, like methane, ethylene, and benzene, are sources of carbon for the CNTs. In the broad sense, this method works on the catalytic decomposition of hydrocarbons on metallic catalysts like Co, Ni, or Fe at high temperatures (~700–900°C) (Mtibe et al. 2018). This generating a high-quality product from a low-temperature energy-efficient synthesis (Soni et al. 2020). There are other, less significant approaches like electrolysis, sonochemical/hydrothermal, pyrolysis, etc., used in practice for CNT synthesis (Soni et al. 2020, Han et al. 2019).

8.4 CARBON NANOTUBES AND GRAPHENE FOR CHEMIRESISTIVE BIOSENSORS

The selection of better transducer/sensing materials is a critical factor in chemiresistive biosensing. Hence, the material selected for this process should possess extreme sensitivity to the minute changes in resistance/conductivity, high available surface areas, and effective surface modification properties (Gong et al. 2013, Chartuprayoon et al. 2015). In this sense, the CNTs and graphene fulfil the aforementioned requirements (Deshmukh et al. 2020). Both materials are known for their superior conducting properties, large available surface areas and hence high adsorption capabilities. The high surface area exhibited by CNTs due to the inherent 1D behaviour induces high aspect ratios (surface area-to-volume ratio), which is an additional advantage. This made research into CNTs explore more transducer elements for biosensing than graphene, though they possess similar properties. However, both materials could be candidate transducer elements for chemiresistive biosensing. Along this line, biosensors with CNTs/graphene as the transducer element are discussed below.

8.4.1 CNT-based Biosensors

CNTs of various types are used as transducer elements in biosensors to detect various microorganisms, explosives, environmental and psychological stressors, diseases, pH changes, human biomarkers, water contaminants, etc. (Table 8.1).

Garcia-Aljaro et al. (2010) have reported a biosensor made up of SWCNTs as the transducer element for the detection of human pathogens, namely the bacterium *Escherichia coli* O157:H7 and the bacteriophage T7 as a virus model. For detection, the transducer elements were functionalized with anti-*E. coli* O157:H7 or anti-T7 antibodies, respectively. The observed detection limits for *E. coli* O157:H7 and T7 bacteriophage were 10^3 CFU mL^{-1} and 10^3 PFU mL^{-1}, respectively. Furthermore, the authors reported that this sensor was simple, operated at room temperature, and carried out real-time sensing. An immunosensor was developed to detect cortisol biomarkers in the saliva, revealing the psychobiologic response to an environmental or psychological stressor (Tlili et al. 2011). The layers of SWCNTs were used as transducer elements and functionalized with the cortisol-3-CMO-NHS ester, to which the monoclonal anti-cortisol antibody was ligated. The detection limit achieved was 1 pg mL^{-1}, making this biosensor very sensititive as well as highly selective. Matta et al. (2016) have developed a biosensor comprising of a MWCNT-embedded SU-8 nanofibre for the detection of cardiac biomarkers, namely myoglobin (Myo), cardiac troponin I (cTn I), and creatine kinase-MB (CK-MB). The transducer element was functionalized with 1-ethyl-3-(3- dimethylaminopropyl) carbodiimide (EDC), and *N*-hydroxysuccinimide (NHS), where the appropriate antibodies were attached using crosslinking chemistry. The detection limits achieved were 6, 20, and 50 fg mL^{-1} for Myo, cTn I, and CK-MB, respectively, with a response time of less than one minute. The authors added that this biosensor could be a candidate for detecting cancer biomarkers at ultra-low concentrations. On the other hand, Fu et al. (2017) developed a biosensor based on semiconducting(sc)-SWNTs/nitrogen-doped (N)-MWCNTs for the detection of the subtype H5N1 DNA sequence of avian influenza virus (AIV). Initially, the oligonucleotide DNA probes were functionalized non-covalently to the CNTs. The H5N1 DNA T biomarker in phosphate buffer (PB) was used as the analyte for sensing measurements. The detection limit was observed to be 2 pM and 20 pM for the sc-SWCNTs and N-MWCNTs, respectively, following an incubation period of 15 mins. The better performance of the sc-SWCNTs could be attributed to

TABLE 8.1
The Biosensing Details for the CNTs and Graphene-Based Biosensors

Transducer Element	Bio-sensor Designed for	Receptor/functional Protein	Biomarker	Antibody	Detection Limit	Response Time	Reference
SWCNTs	Detection of microorganisms (bacteria/ virusES)	1-Pyrene butanoic acid, succinimidyl ester	*Escherichia coli* O157:H7	Anti-*E. coli* O157:H7 antibodies	10^3 CFU mL^{-1}	60 min	Pejcic et al. (2006)
		1-Pyrene butanoic acid, succinimidyl ester	Bacteriophage T7	Anti- T7 antibodies	10^3 PFU mL^{-1}	5 min	
SWCNTs	The hypothalamic-pituitary-adrenal transduced physiological responses due to an environmental or psychological stressor	Cortisol-3-CMO-NHS ester	Salivary cortisol	Monoclonal anti-cortisol antibody	1 pg mL^{-1}	—	García-Aljaro et al. (2010)
MWCNT- embedded SU-8 nanofibre	Acute myocardial infarction	1-Ethyl-3-(3-dimethylaminopropyl) carbodiimide (EDC), and *N*-Hydroxy succinimide (NHS)	Myoglobin, cardiac troponin I, and creatine kinase-MB	—	6, 20, and 50 fg mL^{-1}, respectively	<1 min	Park et al. (2010)
Semiconductor SWCNTs	Detection of AIV H5N1 virus from the biological sample of humans viz., nose swabs and blood	DNA T probe molecules	H5N1 DNA T	DNA T probe molecules	2 pM	15 min	Bandi and Srivastav (2020)
Nitrogen-doped MWCNTs					20 pM	15 min	Bandi and Srivastav (2020)

(Continued)

TABLE 8.1 (CONTINUED)
The Biosensing Details for the CNTs and Graphene-Based Biosensors

Transducer Element	Bio-sensor Designed for	Receptor/functional Protein	Biomarker	Antibody	Detection Limit	Response Time	Reference
Heparin- functionalized SWCNTs	Detection of Dengue virus (DENV)	Heparin	DENV	—	8.4×10^{2} $TCID_{50}$ mL^{-1}	—	Geim and Novoselov (2007)
MWCNTs and ZnO nanofibres	Detection of malaria	—	Histidine-rich protein II (HRP2)	HRP2 antibodies	10 fg mL^{-1}	—	Bandi et al. (2019)
SWCNTs	Mercury detection in biological samples	1-Pyrenebutanoic acid, succinimidyl ester	Divalent mercury and monomethyl mercury	Poly T:poly A duplex	1 nM and 0.5 nM	—	Rao et al. (2018)
SWCNT network	Detection of TNT	Trinitrophenyl (TNP)	TNT	Anti-TNP scAB	0.5 ppb	—	Tasis et al. (2006)
Composite of poly [*N'*-(4-(dodecyl(phenyl) amino) benzylidene)-dithiocarbohydrazone] - poly (DATC) and MWCNTs	Nitrate ion detection in water	—	NO_3^--N	—	14 ppm	~3 min	Novoselov et al. (2004)
SWCNTs	Detection of foodborne pathogenic bacteria	Cobalt porphyrins	Putrescine (butane-1,4-diamine) and cadaverine (pentane-1,5-diamine)	—	—	—	Hummers and Offeman (1958)
Monolayer graphene	Detection of bacteria	Streptavidin	*E. coli* K12 bacteria	Anti-*E. coli* antibodies	12 CFU mL^{-1}	—	Eatemadi et al. (2014)

the modulation of the Schottky-barrier owing to the material's semiconductor behavior (Fu et al. 2017).

A SWCNT nanonetwork-based biosensor was also developed to detect dengue virus (DENV) (Wasik et al. 2017). The intriguing aspect of this biosensor is that the SWCNTs were functionalized with heparin instead of antibodies to achieve the recognition of the virus. The minimum detection limit achieved for DENV was 8.4 $\times 10^2$ $TCID_{50}$ mL^{-1}. The MWCNTs and ZnO nanofibres were investigated to detect a malaria biomarker, i.e., histidine-rich protein II (HRP2) (Paul et al. 2017). The sensor was functionalized with HRP2 antibodies before sensing. The sensor exhibited a sensitivity of 8.29 kΩ g^{-1} mL^{-1} in the wider detection range of 10 fg mL^{-1} to 10 ng mL^{-1} of HRP2. Furthermore, detecting mercury levels from biological samples using chemiresistive biosensors is another intriguing achievement (Gong et al. 2013, Wordofa et al. 2016). The divalent mercury (Hg^{2+}) and monomethyl mercury (CH_3Hg^+) are two forms highly toxic to humans. For detection, the SWCNTs were functionalized with poly T (5′-/5AmMC6/TTT TTT TTT TTT TTT-3′): poly A (5′-AAA AAA AAA AAAAAA-3′) duplex. A simulated saliva solution with different divalent and monomethyl mercury concentrations was spiked onto the sensor for detection. A linear response was observed for both divalent and monomethyl mercury in the ranges of 1–1000 nM and 0.5–500 nM, respectively, with corresponding sensitivities of 15.68% and 24.8% per log (nM), respectively (Wordofa et al. 2016).

In addition to detecting disease-related biomarkers, the CNT-based biosensors were also employed to detect foodborne pathogens, water contaminants, pH changes, etc. A biosensor composed of the SWCNT network as a transducer element was designed to detect 2,4,6-trinitrotoluene (TNT) used in industrial and explosive applications, which is a source of significant groundwater and soil contamination (Park et al. 2010). Firstly, the transducer elements were modified with trinitrophenyl (TNP), further ligated with an antibody, i.e., anti-TNP single-chain antibody (scAB). A 0.5 ppb limit of detection was obtained with the prepared sensor along with high selectivity. Similarly, Mallya et al. (2018) have developed a biosensor for nitrate ion detection in water, where the researchers explored poly-[*N*'-(4-(dodecyl(phenyl)amino) benzylidene)-dithiocarbohydrazone]-poly (DATC) and a MWCNT composite as a transducer element. Thiha et al. (2018) developed a biosensor based on suspended carbon nanowires to detect *Salmonella typhimurium*, a foodborne human-pathogenic bacterium. At the same time, a biosensor composed of SWCNTs functionalized with cobalt porphyrins was developed for the detection of amines to identify meat spoilage (Liu et al. 2015). Decomposition of meat generates biogenic amines which act as biomarkers, namely putrescine (butane-1,4-diamine) and cadaverine (pentane-1,5-diamine).

8.4.2 Graphene-based Biosensors

The use of CNTs for chemiresistive biosensing have frequently been reported, but graphene-based chemiresistive biosensors are relatively less explored. A monolayer graphene-based biosensor was developed by Zhao et al. (2020) to detect *E. coli* K12 bacteria. The sensor was functionalized with the anti-*E. coli* antibodies before the sensing measurement was conducted. In addition, a microchannel integrated into the graphene is an intriguing aspect of this sensor. The lowest detection capability of the sensor was reported to be12 CFU mL^{-1} (Zhao et al. 2020). In addition, the sensor detection of volatile organic compounds from human breath was reported (Liu et al. 2019), although this comes under the category of a gas sensor.

On the other hand, graphene is widely used to detect bacterial and viral pathogens *via* electrochemical- or fluorescent-based biosensing, rather than chemiresistive biosensing (Jiang et al. 2020). It is worth pointing out that chemiresistive gas sensing has been widely explored, with highly efficient results towards various analytes (Bandi and Srivastav 2020). Hence, considering the low-cost preparation, its performance, and its comparative performance relative to CNTs, graphene can also be seen to deliver fascinating sensing properties. Hence, graphene should be explored further for use in chemiresistive biosensors.

8.5 SUMMARY

In summary, the CNTs (in the form of SWCNTs, and MWCNTs) and graphene are emerging as invaluable transducer element materials for use in chemiresistive biosensors. The CNTs in the form of SWCNTs/MWCNTs have been especially highly investigated for the detection of various microorganisms, foodborne pathogens, water contaminants, viral infections, biological sample-based disease identification, etc. At the same time, graphene has also been investigated to detect bacterial pathogens, using chemiresistive biosensing techniques. However, little has been published in this area. Despite this, these materials are used to a significant extent in electrochemical and fluorescent biosensors. As the CNTs and graphene are similar in properties and performance, apart from their morphological dissimilarity, graphene can also be widely explored for a role in chemiresistive biosensing. This concept highlights the straightforward production of transducing elements (graphene) and the ease of device fabrication. The rest of the performance should compete readily with the CNT-based biosensors.

REFERENCES

Bandi S., Srivastav A.K. 2020. Graphene-based chemiresistive gas sensors. *Compr Anal Chem* Elsevier B.V.; 2020:149–73.

Bandi S., Ravuri S., Peshwe D.R., Srivastav A.K. 2019. Graphene from discharged dry cell battery electrodes. *J Hazard Mater* Elsevier B.V.; 366:358–69.

Chartuprayoon N., Zhang M., Bosze W., Choa Y.H., Myung N.V. 2015. One-dimensional nanostructures based bio-detection. *Biosens Bioelectron* Elsevier; 63:432–43.

Deshmukh M.A., Jeon J.Y., Ha T.J. 2020. Carbon nanotubes: An effective platform for biomedical electronics. *Biosens Bioelectron* Elsevier B.V.; 150:111919.

Eatemadi A., Daraee H., Karimkhanloo H., Kouhi M., Zarghami N., Akbarzadeh A., et al. 2014. Carbon nanotubes: Properties, synthesis, purification, and medical applications. *Nanoscale Res Lett* 9:393.

Fu Y., Romay V., Liu Y., Ibarlucea B., Baraban L., Khavrus V., et al. 2017. Chemiresistive biosensors based on carbon nanotubes for label-free detection of DNA sequences derived from avian influenza virus H5N1. *Sensors Actuators, B Chem* Elsevier B.V.; 249:691–9.

García-Aljaro C., Cella L.N., Shirale D.J., Park M., Muñoz F.J., Yates M.V., et al. 2010. Carbon nanotubes-based chemiresistive biosensors for detection of microorganisms. *Biosens Bioelectron* 26:1437–41.

Geim A.K., Novoselov K.S. 2007. The rise of graphene. *Nat Mater* 6:183–91.

Gong J.L., Sarkar T., Badhulika S., Mulchandani A. 2013. Label-free chemiresistive biosensor for mercury (II) based on single-walled carbon nanotubes and structure-switching DNA. *Appl Phys Lett* 102:2012–5.

Han T., Nag A., Chandra Mukhopadhyay S., Xu Y. 2019. Carbon nanotubes and its gas-sensing applications: A review. *Sens Actuators, A Phys* Elsevier B.V.; 291:107–43.

Hummers W.S., Offeman R.E. 1958. Preparation of graphitic oxide. *J Am Chem Soc* 80:1339.

Jiang Z., Feng B., Xu J., Qing T., Zhang P., Qing Z. 2020. Graphene biosensors for bacterial and viral pathogens. *Biosens Bioelectron* Elsevier B.V.; 166:112471.

Liu B., Huang Y., Kam K.W., Cheung W.F., Zhao N., Zheng B. 2019. Functionalized graphene-based chemiresistive electronic nose for discrimination of disease-related volatile organic compounds. *Biosens Bioelectron X* Elsevier; 1:100016.

Liu S.F., Petty A.R., Sazama G.T., Swager T.M. 2015. Single-Walled carbon nanotube/metalloporphyrin composites for the chemiresistive detection of amines and meat spoilage. *Angew Chemie: Int Ed* 54:6554–7.

Mallya A.N., Ramamurthy P.C. 2018. Design and fabrication of a highly stable polymer carbon nanotube nanocomposite chemiresistive sensor for nitrate ion detection in water. *ECS J Solid State Sci Technol* 7:Q3054–64.

Matta D.P., Tripathy S., Krishna Vanjari S.R., Sharma C.S., Singh S.G. 2016. An ultrasensitive label free nanobiosensor platform for the detection of cardiac biomarkers. *Biomed Microdevices* 18:1–10.

Mtibe A., Mokhothu T.H., John M.J., Mokhena T.C., Mochane M.J. 2018. Fabrication and characterization of various engineered nanomaterials. *Handb Nanomater Ind Appl* Elsevier; 2018:151–71.

Novoselov K.S, Geim A.K., Morozov S.V., Jiang D., Zhang Y., Dubonos S.V., et al. 2004. Electric field effect in atomically thin carbon films. *Science (80-)* 306:666–70.

Park M., Cella L.N., Chen W., Myung N.V., Mulchandani A. 2010. Carbon nanotubes-based chemiresistive immunosensor for small molecules: Detection of nitroaromatic explosives. *Biosens Bioelectron* 26:1297–301.

Paul B., Panigrahi A.K., Singh V., Singh S.G. 2017. A multi-walled carbon nanotube-zinc oxide nanofiber based flexible chemiresistive biosensor for malaria biomarker detection. *Analyst* 142:2128–35.

Pejcic B., Marco R. de , Parkinson G. 2006. The role of biosensors in the detection of emerging infectious diseases. *Analyst* 131:1079.

Rao R., Pint C.L., Islam A.E., Weatherup R.S., Hofmann S., Meshot E.R., et al. 2018. Carbon nanotubes and related nanomaterials: Critical advances and challenges for synthesis toward mainstream commercial applications. *ACS Nano* 12:11756–84.

Soni S.K., Thomas B., Kar V.R. 2020. A comprehensive review on CNTs and CNT-reinforced composites: Syntheses, characteristics and applications. *Mater Today Commun* Elsevier; 25:101546.

Tasis D., Tagmatarchis N., Bianco A., Prato M. 2006. Chemistry of carbon nanotubes. *Chem Rev* 106:1105–36.

Thiha A., Ibrahim F., Muniandy S., Dinshaw I.J., Teh S.J., Thong K.L., et al. 2018. All-carbon suspended nanowire sensors as a rapid highly-sensitive label-free chemiresistive biosensing platform. *Biosens Bioelectron* Elsevier B.V.; 107:145–52.

Tlili C., Myung N.V., Shetty V., Mulchandani A. 2011. Label-free, chemiresistor immunosensor for stress biomarker cortisol in saliva. *Biosens Bioelectron* Elsevier B.V.; 26:4382–6.

Wasik D., Mulchandani A., Yates M.V. 2017. A heparin-functionalized carbon nanotube-based affinity biosensor for dengue virus. *Biosens Bioelectron* Elsevier; 91:811–6.

Wordofa D.N., Ramnani P., Tran T.T., Mulchandani A. 2016. An oligonucleotide-functionalized carbon nanotube chemiresistor for sensitive detection of mercury in saliva. *Analyst* Royal Society of Chemistry; 141:2756–60.

Zhao W., Xing Y., Lin Y., Gao Y., Wu M., Xu J. 2020. Monolayer graphene chemiresistive biosensor for rapid bacteria detection in a microchannel. *Sensors and Actuators Reports* Elsevier B.V.; 2:100004.

9 Green Synthesis of Graphene and Graphene Oxide and Their Use as Antimicrobial Agents

Roberta Bussamara, Nathália M. Galdino, Andrea A. H. da Rocha, and Jackson D. Scholten

CONTENTS

9.1 INTRODUCTION

Nanomaterials are a class of materials on the nanometer scale with a unique characteristic: their size can modulate their properties. For this reason, these materials have been used for several purposes with greater effectiveness, especially carbon-based nanomaterials. Carbon-based nanomaterials stand out for their biocompatibility and simple precursors. Carbon nanomaterials include nanodiamonds, fullerenes, carbon nanohorns, carbon nano-onions, single-walled carbon nanotubes, multi-walled carbon nanotubes, unzipped carbon nanotubes, carbon quantum dots, and graphene and graphene oxide.

Graphene has attracted attention from the scientific community since its structure was determined in 2004 by Novoselov et al. (2004): the pristine (pure, unoxidized)

DOI: 10.1201/9781003110781-9

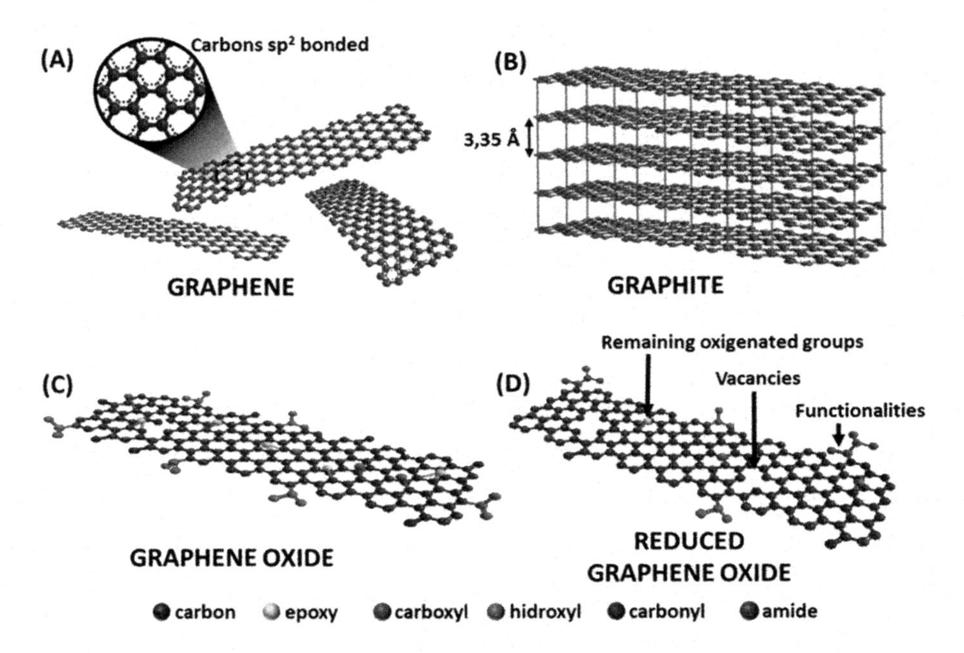

FIGURE 9.1 Structure of pristine graphene sheets (A), graphite (B), graphene oxide (C) and reduced graphene oxide (D). The hydrogen atoms are hidden.

or intrinsic graphene layer is a one-atom-thickness of sp2-bonded carbon atoms in a honeycomb crystal lattice (Figure 9.1A). Graphite is a regular precursor of graphene, where the graphite is composed of several graphene sheets interacting with each other by van der Waals forces (Figure 9.1B). The significant advantage of graphene-based materials over other carbon-based materials is the higher specific surface area, and their optical, mechanical, electrical, and electrochemical properties.

Graphene oxide and reduced graphene oxide are derivatives of graphene. They are used primarily in the biomedical field due to their biocompatibility and the possibility of immobilizing biomolecules onto them. Graphene oxide is a graphene sheet with oxide groups randomly distributed as epoxy or hydroxyl in the basal plane and carboxyl or carbonyl at the edges (Figure 9.1C). The C/O ratio and the percentage of each oxide group depend on the synthesis conditions. Graphene oxide can be reduced to reduced graphene oxide, which retains some oxygenated groups, mainly at the edges, and has some structural defects on the basal plane (Figure 9.1D). The reduction process can be thermal, electrochemical, chemical, or biological. Hydrazine, hydroxylamine, hydroquinone, ascorbic acid, glucose, sodium borohydride, pyrrole and alkaline solutions are all examples of chemical reducing agents which have been used (Zhu et al. 2010, Lim et al. 2018).

Graphene is a robust material with versatile properties, making it suitable for several applications (Figure 9.2). Furthermore, graphene oxide offers countless possibilities for chemical modification, increasing the efficiency and specificity according to the required purpose.

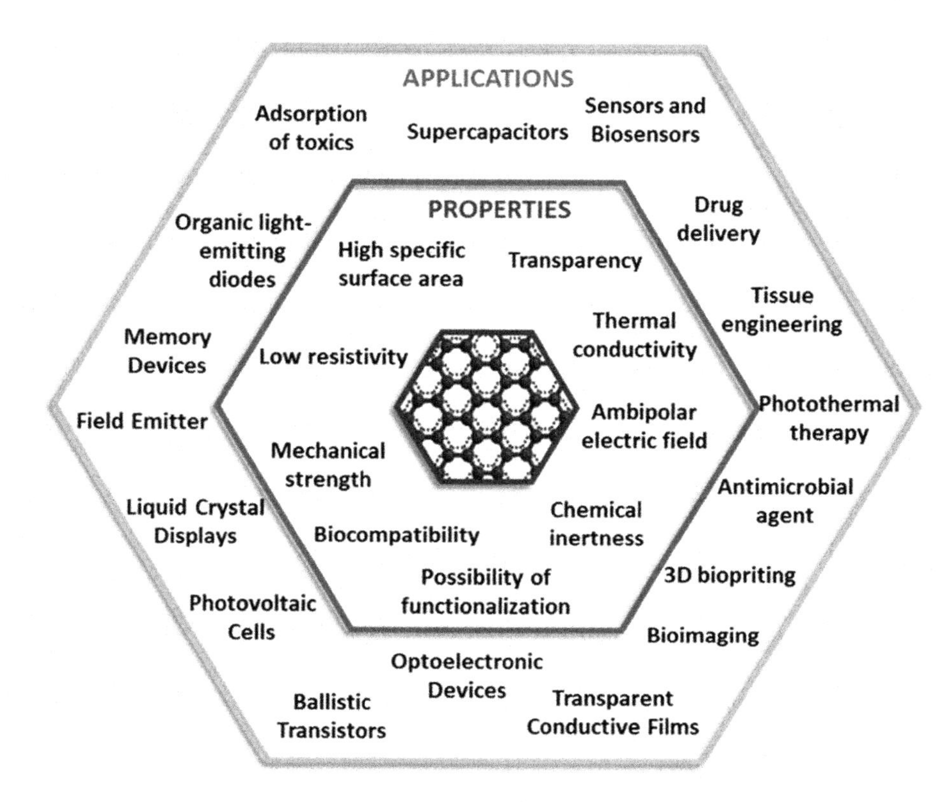

FIGURE 9.2 Properties and applications of graphene-based materials.

Recently, the green synthesis of graphene-based materials has been promoted as an environmentally friendly, cost-effective, low-energy and relatively simple approach to large-scale production of high-performance materials in several areas. Gurunathan et al. (2013) used *Escherichia coli* biomass to synthesize reduced graphene oxide with antitumour properties. The *E. coli* contain proteins, which were used as a reducing agent for graphene oxide. Even though pristine graphene can trigger apoptosis by the mitochondrial pathway, the biologically reduced graphene oxide decreased the viability of the MCF-7 human breast cancer cell by 64%, compared with pure graphene. In addition, the biologically reduced graphene oxide showed an increase in tumour cell mortality of approximately 50%, which makes it a promising material in cancer nanotherapy (Gurunathan et al. 2013a).

Additionally, there are many examples of graphene oxide combined with other materials to obtain nanocomposites with high-performance and synergistic properties. Reduced graphene oxide with iron nanoparticles (RGO/Fe NPs) was obtained by one-step bio-reduction using eucalyptus leaf extract. In addition to the reduction function, the plant extract also capped the surface of the RGO/Fe NP hybrid material that helped stabilize the nanostructure, as observed by infrared spectroscopy. The RGO/Fe NP hybrid material could remove up to 99.7% of methylene blue dye from aqueous solutions. The high dispersibility and stabilization of the Fe NPs onto RGO make them a promising green material for wastewater remediation (Weng et al. 2018).

In another study, Keerthi et al. (2018) developed an electrode coated with reduced graphene oxide/Prussian blue microcubes obtained by green synthesis mediated by a mushroom extract. The screen-printed sensor was applied to detect the prophylactic drug 1, 2-dimethyl-5-nitroimidazole, which has carcinogenic and mutagenic properties. The amperometric sensor ranged linearly from 0.02 μM to 1360.1 μM at pH 5 and was applied to drug-spiked milk and egg samples (Keerthi et al. 2018).

Due to the enormous versatility of green-synthesized graphene-based materials, this chapter will present some green approaches for graphene and graphene oxide nanomaterials and their application as antimicrobial agents. For comparison, initially, some properties of graphene and graphene oxide nanomaterials and the traditional primary synthesis of these materials will be presented.

9.2 GRAPHENE AND GRAPHENE OXIDE CHARACTERISTICS AND PROPERTIES

Pristine graphene is a unique material with the highest theoretical specific surface area (2630 m^2 g^{-1}) among the carbon materials. Furthermore, it exhibits high values of thermal conductivity (~5000 W m^{-1} K^{-1}), Young's modulus (~1 TPa), fracture strength (~130 GPa), intrinsic mobility (~200000 cm^2 v^{-1} s^{-1}), and optical transmittance (~97.7%) (Shah et al. 2015). The properties of graphene-based materials are changeable according to the number of sheets, modification of the structure, and functionalization.

Depending on the synthesis method, graphene can be single-layer, bi-layer, or few-layer (three to ten layers). Whenever the graphene has more than one layer, the sheets will interact through the π orbitals perpendicular to the plane by hexagonal (or AA), Bernal (or AB) or rhombohedral (or ABC) stacking. The number of layers can be estimated by optical microscopy on a Si substrate, transmission electron microscopy, or Raman spectroscopy. Raman spectroscopy is suitable for determining the thickness of graphene from the position, intensity, width, and shape of the G and 2D bands (Choi et al. 2010).

The modifications of the structure of graphene are called defects and might be intentional or a consequence of the synthetic method. The defects include sp^3 carbons (on the edges, cracks, or functionalities), topological defects (as pentagons and heptagons), adatoms or vacancies, or impurities, among others.

Since graphene has carbon atoms bonded with a delocalized network of electrons, graphene-based materials present unique electronic structures. The Brillouin zone and the quantum Hall effect presented by graphene are thoroughly explained in the literature (Choi et al. 2010). These properties explain why graphene is a semiconductor with modulated gap energy, even acting as a gapless semiconductor.

Graphene has an ambi-polar electric field, and consequentially the charge carriers might be electrons or holes at room temperature, depending on the nature of the gate voltage. Additionally, the mobility of the carriers depends weakly on the temperature, which makes the graphene-based materials excellent conductors, even at room temperature. Because the adsorbed molecules modify the local carrier's concentration, and graphene is a low-noise material, single-layer graphene may be used as a sensor. Schedin et al. have reported that single-layer graphene can detect the adsorption of single gas molecules by changing resistance (Schedin et al. 2007). On the other hand, the thermal conductivity occurs by phonon transport because pristine graphene has a low carrier density which varies with the length of the sheets.

Another advantage is the high mechanical strength of defect-free graphene, expressed by high values of Young's modulus and greater fracture strength. The increase in the defects, including chemical functionalization, changes the mechanical properties. Whereas the graphene oxide monolayers have lower Young's modulus and fracture strength than the pristine graphene, the mechanical properties might be improved by increasing the number of layers or cross-linking agents between the graphene sheets (Zhu et al. 2010). Graphene is commonly used in nanocomposites to improve mechanical properties. In addition, graphene single and bi-layers have incredibly high transparency to ultra-violet to infra-red light. As expected, the opacity increases linearly with the number of layers.

The main challenge to working with pristine graphene is solubilization; in this case, graphene oxide's hydrophilic oxygen functional groups help solubilize graphene in several solvents. Although the pristine graphene is superhydrophobic, graphene oxide disperses quickly by sonication in water.

The modification of the electronic structure is a significant drawback. Some modifications, such as covalent functionalization, can change the properties of graphene materials. Covalent or non-covalent functionalization is usually performed using graphene oxide as a precursor, although it can take place directly on the carbons of graphene.

9.3 TRADITIONAL SYNTHESIS OF GRAPHENE AND GRAPHENE OXIDE NANOMATERIALS

The synthesis of graphene and graphene oxide nanomaterials are performed by different methodologies, depending on the required quantity and desired application. Each of these methodologies is going to produce graphene-based material with a particular structure and properties. Figure 9.3 summarizes some traditional strategies of synthesis of graphene and graphene oxide nanomaterials presented in this section of Chapter 9.

There are two main approaches to graphene synthesis. The first one begins with graphite and separate graphene layers, such as mechanical or chemical exfoliation and chemical synthesis. In contrast, different carbon sources are used to grow the graphene sheets in chemical vapour deposition and epitaxial growth.

9.3.1 Mechanical or Chemical Exfoliation

Since many graphene sheets, bonded by van der Waals force, form graphite, it is feasible to extract one layer by breaking this force. The energy used to break the interaction between sheets might be mechanical or chemical.

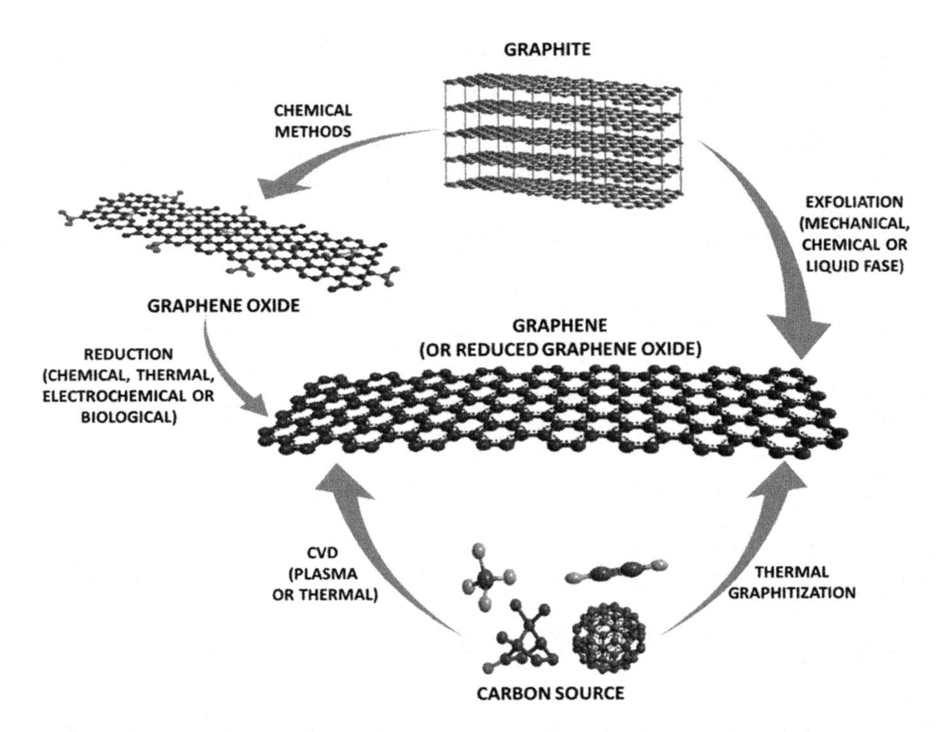

FIGURE 9.3 Most common traditional methods of graphene synthesis that use graphite or organic molecules as a precursor.

Chemical cleavage was the first approach to extracting graphene directly from graphite, used by Viculis et al., whose graphene was obtained by intercalating graphite sheets with potassium metal by exfoliation with ethanol. Unfortunately, the exfoliation led to the formation of nanoscrolls with 40 ± 15 layers in each sheet (Viculis et al. 2003). The few-layers and single-layer graphene were finally obtained using a mechanical approach by Novoselov in 2004. In Novoselov's study, the highly oriented pyrolytic graphite was dried in oxygen plasma, placed in a photoresist, and baked. The graphite was peeled using scotch tape to obtain the graphene sheets attached to the photoresist. The graphene sheet was released in acetone and transferred to a Si substrate (Novoselov et al. 2004).

Furthermore, the liquid phase exfoliation allowed the large-scale generation of products. However, the oxidation and reduction steps resulted in structural defects and damage to the electrical properties. Under high-intensity ultrasound, the graphite layers are separated into organic solvents such as *N*-methyl-2-pyrrolidone or dimethylformamide (Zhu et al. 2010).

This strategy can be carried out by electric field, ultra-sonication, the transfer printing technique, or scotch tape. These methods are suitable for producing high-quality graphene, especially for electrical purposes, although, in general, the flakes obtained are small and with low yields.

9.3.2 Epitaxial Growth

Thermal graphitization or epitaxial growth on a SiC surface is the most popular method for synthesizing graphene with few layers, especially for semiconductor applications. In a typical experiment, the graphene is formed on the surface plane (001) of a 6H-SiC single crystal which is heated to extremely high temperatures (1250–1450 °C) for a short period (1–20 min). This method can produce graphene with a large size and few layers compared to the others, but it is not used on a large scale because there are still problems with control of the homogeneity. The thickness of graphene seems to be controlled by the surface plane of the substrate and the parameters (pressure, temperature, and heating rate) (Choi et al. 2010).

9.3.3 Chemical Vapor Deposition

The chemical vapour deposition (CVD) procedure deposits a gaseous carbon precursor onto a substrate, carried by thermal or plasma energy. The first publication using thermal CVD is from 2006, where camphor was used as the precursor of plane few-layers graphene sheets on Ni foil. In this particular study, the camphor was evaporated at 180°C and pyrolyzed from 700 to 850°C in a CVD furnace with an argon atmosphere. Using a sharp blade, the graphene sheets were scraped from the Ni fold after cooling to room temperature (Somani et al. 2006).

Thermal CVD has been producing graphene with reproducibility and desired properties, especially for photovoltaic and flexible electronics. Parameters, like the precursor, gas pressure, cooling rate, the substrate used for the synthesis, and the transfer for other substrates, might change the quality of the graphene produced. Other CVD procedures have been explored to minimize the substrate effect, such as plasma-enhanced chemical vapour deposition (PECVD).

9.3.4 Chemical Methods

The chemical methods produce graphene oxide from graphite as a result of intercalation, oxidation, and exfoliation steps. The intercalation agents might be atoms, ions or molecules, and the exfoliation of graphene sheets is usually by ultrasonic vibration or thermal shock.

There are four generations of chemical synthesis of graphene oxide. During these generations, the intercalations and oxidation reagents are modified. Brodie was the first to report the oxidation of slurry graphite by adding potassium chlorate and concentrated nitric acid in 1859 (Brodie 1859). After 40 years, Staudenmaier improved this method by using a mixture of acids (HNO_3 and H_2SO_4) (Staudenmaier 1898). Afterwards, Hummers and Offeman developed an alternative strategy to synthesize graphene, which has been widely used up to the present time (Hummers and Offeman 1958).

Hummers and Offeman firstly published 1958 this method as the preparation of graphitic oxide. In the original method, the graphite was oxidized to the graphitic oxide by mixing the graphitic powder with concentrated sulfuric acid, sodium

TABLE 9.1
Overview of Main Advantages and Drawbacks in Traditional Methods of Synthesis of Graphene and Graphene Oxide Materials

Method	Advantages	Drawbacks
Exfoliation	High-quality graphene Single-layer graphene	Low yields Small flakes
Thermal graphitization	Large size of graphene Few-layer graphene	High temperatures applied Low homogeneity
CVD	Graphene with high crystallinity Simple precursors	Substrate effect Number of layers Problems with large-scale production
Chemical exfoliation	Large-scale production Water-soluble graphene oxide	Use of strong oxidants and concentrated acids, some of which are toxic. Time consuming

nitrate and potassium permanganate for a few hours (Hummers and Offeman 1958). However, the toxic gas generation (NO_2 and N_2O_4) and the residual nitrate are drawbacks of the original Hummers method. Therefore, many modified Hummer's methods have been published to decrease the toxicity, control the number of layers, and increase the O/C ratio. Reduced graphene oxide might be achieved after the reduction of graphene oxide synthesized by the chemical method.

Table 9.1 summarizes the main advantages and drawbacks of the traditional synthesis methods of graphene and graphene oxide explained above.

The biological reduction of graphene oxide is one of the green synthesis strategies for obtaining graphene for several purposes. Other less usual techniques for graphene synthesis include carbon nanotubes opening, arc-discharge method, substrate-free gas-phase synthesis, and the template route (Lee et al. 2019).

9.4 GREEN APPROACHES TO GRAPHENE AND GRAPHENE OXIDE MATERIALS

As described in the previous discussions, it is possible to notice several application possibilities of graphene and graphene oxide materials. However, the use of toxic substances, the need for sophisticated instrumentation, and the time-consuming and high-cost methods necessary for large-scale production make the synthesis process of these materials a real challenge (Cherian et al. 2019).

The preparation of graphene and graphene oxide materials by traditional methods may present some drawbacks, such as high temperatures, time-consuming methods, low material homogeneity, substrate effects, problems with large-scale production, and the use of strong oxidants and concentrated acids, among others. Consequently, medical applications can be compromised due to problems with biocompatibility, toxicity, and high cost (Bhardwaj et al. 2020).

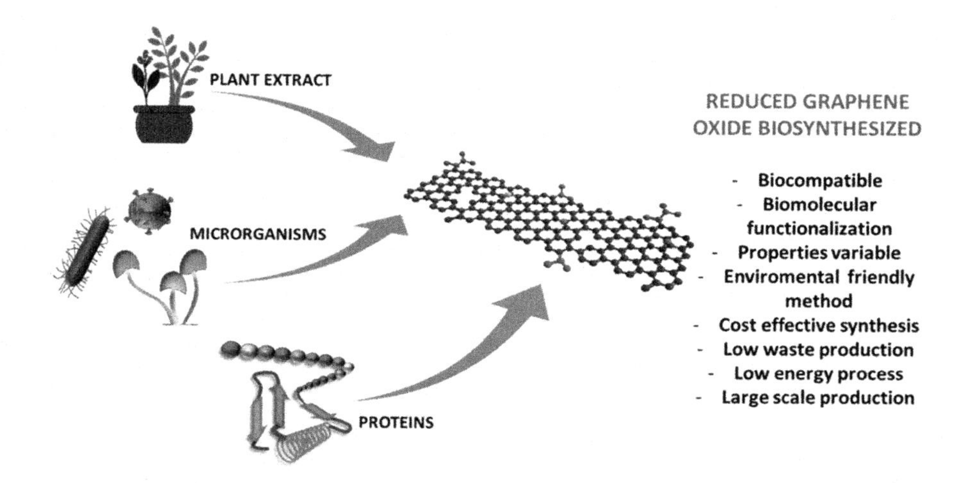

FIGURE 9.4 Biosynthetic approaches to obtain reduced graphene oxide and the advantages of these methods.

Green nanotechnology has emerged as an alternative to minimize these drawbacks. Usually, green nanotechnology uses organisms (bacteria, fungi, plants, viruses) in biological routes, biomolecules isolated from them, waste materials, and environment-friendly physical methods for nanomaterial production. Compared to traditional methods, this approach is more environmentally friendly, non toxic and cost effective (Cruz et al. 2020, Sadhasivam et al. 2020). In addition, these green nanomaterials show essential characteristics that positively impact health and the environment, such as biocompatibility, biodegradability, and low waste production.

Figure 9.4 shows three green approaches to obtain reduced graphene oxide materials, and their advantages.

Graphene and graphene oxide produced by the green route can offer many possibilities for application in the medical and pharmaceutical fields. It is due to their capacity for functionalization of the material's surface through interaction with biomolecules, which results in biomaterials with important characteristics necessary for the biological systems. Nanoparticles may be combined with these materials to attain new properties, making them good candidates for antimicrobial, biomedical, and food-related applications (Azizi et al. 2020).

9.4.1 Green Synthesis Mediated by Plant Extracts

Graphene oxide reduction, mediated by plant extracts, is a promising green methodology for producing graphene on a large scale. The plant extract is an eco-friendly and widely available alternative to reducing graphene oxide, using low energy and biocompatible processes.

More than one reducing agent is involved in this process since the plant extract combines several natural antioxidants, such as amino acids, proteins, vitamins, glycosides, hormones, and other bioactive substances. A likely mechanism is the major component of plant extracts that is the (poly)phenolic compounds. The phenolic compounds are easily oxidized to the corresponding quinone form by reacting with graphene oxide-oxygen groups (Thakur and Karak 2015). The main advantage of this method is the stability of the dispersion of plant extract-reduced graphene oxide in water, possibly favored by the biomolecular functionalization (Ismail 2019, Mhamane et al. 2011).

In a typical experiment of graphene oxide reduction by a plant extract, the graphene oxide aqueous solution is initially sonicated to exfoliate the graphene oxide sheets. Following this, the graphene oxide is mixed with the plant extract. The mixture is stirred or refluxed, and the colour reflects the reduction, changing from brown to black. The reduced graphene oxide sheets are finally isolated from the plant extract and unreduced graphene oxide by centrifugation (Ismail 2019). The quality of graphene-based material obtained depends on the starting graphene oxide, reaction conditions, and the chosen plant. The compounds and their concentration in plant extracts vary with the species.

Some examples of different plant extract syntheses are presented in Table 9.2. Because the reaction conditions and the type of plant extract are distinct, the reduced graphene oxide produced has a unique structure and properties. Notice that the structure of graphene, and its degree of disorder, are commonly expressed by the region between the D and G bands on Raman spectroscopy.

Reduced graphene oxide, obtained by plant extracts, are applied in several areas, such as in dye removal (Weng et al. 2018), transparent, highly conductive films (Chamoli et al. 2016), and in the biomedical field (Lingaraju et al. 2019). The green synthesis of reduced graphene oxide has been published for an extensive range of plant species in recent years. The reaction conditions and the reduction agents' composition and concentration still need to be explored to produce materials with suitable properties.

9.4.2 Green Synthesis Mediated by Proteins

Natural biopolymers, such as proteins, are used to produce nanomaterials because they are a low-cost and environmentally friendly approach. Proteins can act as reducing, capping, and stabilizing agents. Their structural characteristics, based on the amino acid sequence, provide a highly reactive biomolecule capable of interacting with metal ions and functional groups from the materials (Rana et al. 2020). Nanomaterial production mediated by isolated proteins may present an advantage compared to microorganisms and plant extracts. It is a reproducible methodology due to the controllable protein sequences and structures (Rakhimol et al. 2020).

Frequently, the synthesis of graphene nanomaterials mediated by proteins are simple and require no complex instrumentation. The first step of graphene oxide reduction by proteins involves preparing a graphene oxide aqueous solution. Afterwards, the graphene oxide solution is mixed with powder or liquid protein at a specific pH, for a specific incubation time, and at a controlled temperature. Then, the solution is filtered and dried to obtain the reduced graphene oxide sheets.

TABLE 9.2
Reaction Conditions of Reduced Graphene Oxide Synthesis Mediated by Different Plant Extracts and its Structural Characteristics and Properties

Plant Extract	Reaction Conditions[a]	I_D/I_G[b]	Biosynthesized Reduced Graphene Oxide Advantages	Reference
Cyperus difformis and aquatic macrophytes (*Potamogeton pectinatus, Ceratophyllum demersum, Lemna gibba*)	Water, 90 °C, 24 h [GO] = 0.2 mg mL^{-1} [PE] = 3.1 mg mL^{-1}	0.97	Dispersion stability in water up to 24 h	Mhamane et al. (2011)
Artemisinin (active ingredient of *Artemisia annua*)	Ethanol, 95 °C, 24 h [GO] = 1 mg mL^{-1} [PE] = 2–10 mg mL^{-1}	1.16–1.32	The increase in artemisinin concentration results in smaller sp^2 domains	Hou et al. (2018)
Tribulus terrestris or mint extracts	Water, 180 °C, 12 h [GO] = 1.5 mg mL^{-1} [PE] = 5 mg mL^{-1}	0.98–1.01	Increase of peak current of carbon paste electrode on cyclic voltammetry measurement.	Khojasteh et al. (2019)
Phoenix dactylifera, Cannabis sativa, Citrus × limon, *Punica granatum*	Water/methanol, 80 °C, 24 h [GO] = 0.9 mg mL^{-1} [PE] = 0.9 mg mL^{-1}	1.27–1.38	Increase in the wavelength absorption due to additional delocalization.	Ousaleh et al. (2020)
Ficus religiosa, Mangifera indica and *Polyalthia longifolia* leaves	Water, 50 °C, 24 h [GO] = 0.1 mg mL^{-1} [PE] = 25 mg mL^{-1}	1.12–1.21	Transmittance of spray films near 57% at 550 nm.	Chamoli et al. (2016)
Euphorbia heterophylla leaves	Water, 95 °C, 12 h [GO] = 0.4 mg mL^{-1} [PE] = 0.1 mg mL^{-1}	1.01	Significant cytotoxic activity against cancer cell lines.	Lingaraju et al. (2019)
Artemisia vulgaris	Water, 90 °C, 6 or 12 h [GO] = 0.5 mg mL^{-1}	0.93–0.95	Exhibit luminescent properties that are dependent on the excitation wavelength.	Chettri et al. (2016)

[a] The concentrations of graphene oxide ([GO]) and plant extract ([PE]) were calculated considering the information provided in the experimental section. [b]Values provided by the authors according to Raman spectroscopy.

It is important to highlight as advantages that proteins can reduce the toxicity of nanomaterials and enhance their cellular internalization as soon as they functionalize the graphene oxide and thus facilitate the interaction with other cell biomolecules (Somu and Paul 2019). These biological routes can provide materials such as graphene oxide-based composites with amphiphilic nature and high mechanical strength (Kim et al. 2020). Some examples of different green synthesis processes mediated by proteins are presented in Table 9.3.

9.4.3 Green Synthesis Mediated by Microorganisms

Microorganisms can produce natural substances, such as saccharides, proteins, amino acids, and nucleic acids, that can promote the reduction of graphene oxide (Wang et al. 2021).

Microorganisms such as *Shewanella* spp. (Salas et al. 2010), *Escherichia coli*, (Gurunathan et al. 2013b), *Azotobacter chroococcum* (Chen et al. 2017), *Ganoderma* spp (Gurunathan et al. 2014) and some yeasts (Khanra et al. 2012), have been used as reducing agents in graphene oxide reduction.

In general, the experimental methodology that involves reducing graphene oxide in the presence of microorganisms begins with microorganism cultivation, which requires a controlled incubation time, agitation, and temperature. Subsequently, to reduce the graphene oxide, the graphene oxide solution is added to the microorganism biomass or the extract from the culture medium. This step occurs under agitation under controlled temperature and time. Finally, the mixture is centrifuged or filtered to separate the nanomaterials.

Some mechanisms that explain the reduction of graphene oxide by microorganisms are proposed in the literature (Vargas et al. 2019, Agarwal and Zetterlund 2021). These mechanisms suggest that graphene oxide reduction by microorganisms can be mediated by the microbial respiration process (direct electron transference from cell to graphene oxide material mediated by inner/outer membrane protein complexes); electroactive metabolic substances produced by a microorganism; intracellular redox components released by microorganisms by cell lysis, or by chemical oxidation (generation of reactive oxygen species, ROS) (Agarwal and Zetterlund 2021).

Examples of graphene oxide reduction mediated by microorganisms are presented in Table 9.3.

9.5 ANTIMICROBIAL APPLICATIONS OF GRAPHENE AND GRAPHENE OXIDE NANOMATERIALS

The development of new materials with antimicrobial properties is an important goal of nanotechnology due to the need for new materials that can effectively replace commonly used drugs. Antimicrobial resistance is a real problem that can significantly affect the health of the world population (Nichols and Chen 2020).

Thus, the use of nanomaterials in antimicrobial applications has been of great interest in nanotechnology. As discussed in previous sections, graphene nanomaterials may present an antimicrobial and antioxidant capacity. In that perspective,

TABLE 9.3
Examples of Graphene Oxide Reduction Processes Mediated by Biological Resources (Isolated Proteins and Microorganisms)

Biological Resource	Reaction Conditions[a]	I_D/I_G[b]	Biosynthesized Reduced Graphene Oxide Advantages	Reference
Mussel foot proteins (Mfp) secreted by *Mytilus galloprovincialis*	Acetic acid, on ice, 24 h [GO] = 0.1 mg mL^{-1} [protein] = 0.02 mg mL^{-1}	—	Films have low stiffness, high tensile strength, and ultra-high toughness. Simple green synthesis process. Attractive for a variety of applications.	Kim et al. (2020)
Isolated soy protein	Water, 70 °C, 24 h [GO] = 0.1 wt% [protein] = solid mass ratios of 10:1, 50:1, 100:1 for protein/graphene oxide	1.83–2.30	Easy, one-step approach. Protein/graphene oxide material showed good biocompatibility and exhibited much better photothermal effects than those of pure graphene oxide.	Song et al. (2016)
Casein	Water, 90 °C, 7 h [GO] = 0.1 mg mL^{-1} [protein] = 10 mg mL^{-1}	—	Facile, environmentally friendly, and one-step synthetic method. Casein acts as a reducing and stabilizing agent.	Maddinedi et al. (2014)
Bovine serum albumin (BSA)	Water, 5590 °C, 3-24 h [GO] = 0.1 mg mL^{-1} [protein] = 50 mg mL^{-1}	—	Simple green chemistry route for the decoration and reduction of GO. Possible adhesion of cells onto the surfaces of graphene nanosheets decorated with extracellular matrix proteins. It performed the assembly/co-assembly of pre-synthesized nanoparticles with distinctive sizes, compositions, shapes, and properties. Protein molecules on the nanosheets might be employed to template the *in-situ* growth of metal clusters.	Liu et al. (2010)

(*Continued*)

TABLE 9.3 (CONTINUED)
Examples of Graphene Oxide Reduction Processes Mediated by Biological Resources (Isolated Proteins and Microorganisms)

Biological Resource	Reaction Conditions[a]	I_D/I_G[b]	Biosynthesized Reduced Graphene Oxide Advantages	Reference
Extracellular polymeric substances (EPS) extracted from *Bacillus* sp.	Water, 40 °C, 24 h [GO] = 0.3 mg mL^{-1} [EPS] = 0.1 mg mL^{-1}	1.02	Electron-rich protein from the extracellular polymeric substances reacted easily with graphene oxide functional groups. Extracellular polymeric substances are eco-friendly and non-toxic substances.	Wang et al. (2021)
Escherichia coli biomass	Water, 37 °C, 72 h [GO] = 0.5 mg mL^{-1} *E. coli* biomass = 200 mg mL^{-1}	2.60	Cost-effective, environmentally friendly, and simple approach. *E. coli* acts as a reducing and stabilizing agent that could produce water-dispersible graphene.	Gurunathan et al. (2013b)
Azotobacter chroococcum culture	Water, room temperature, 72 h [GO] = 1 mg mL^{-1} [*A. chroococcum* culture] = 1 mg mL^{-1}	—	No use of toxic chemical agents. Decrease in the agglomeration of reduced graphene oxide.	Chen et al. (2017)
Bacterial consortium biomass	Water, 20–25 °C, 72 h [GO] = 0.4 mg mL^{-1} [bacterial biomass] = 30 mg mL^{-1}	—	High ability to induce the reduction of graphene oxide and an environmentally friendly approach.	Vargas et al. (2019)
Baker's yeast	Water, 35–40 °C, 72 h [GO] = 0.5 mg mL^{-1} [yeast] = 1 mg mL^{-1}	1.34	Environmentally friendly approach. Easy product isolation process, producing water-dispersible graphene.	Khanra et al. (2012)

[a] The concentrations of graphene oxide ([GO]) and biological resources were calculated considering the information provided in the experimental section.

many production routes have been developed, mainly green approaches, to increase these properties (Hassanein et al. 2019). Bioactive compounds from microorganisms, plants, and other living organisms can promote biomolecular functionalization of graphene oxide and increase their antimicrobial effects.

The antimicrobial activity of graphene and graphene oxide nanomaterials depends on, among other factors, the material interaction with the microbial cell (Anand et al. 2019). The graphene cytotoxicity mechanism is widely discussed in the literature. Some assumptions that aim to explain the antimicrobial activity of graphene are oxidative stress induction, protein dysfunction, membrane damage, and transcriptional arrest (Rojas et al. 2020).

The green graphene-based materials are widely used as antimicrobial agents, generally combined with metal nanoparticles (NPs). Silver has presented excellent antimicrobial activity against Gram-positive and Gram-negative bacteria, and a synergistic effect was observed when combined with graphene. Song and colleagues have studied the simultaneous reduction of Ag NPs on reduced graphene oxide (RGO) surfaces mediated by the bacteria *Shewanella oneidensis* MR-1. The nanocomposite was applied as an antibacterial agent against *E. coli*. The Ag/RGO material showed a synergistic effect that resulted in the lowest viability of *E. coli* ($<$ 1%) within 15 min of incubation. Although Ag/RGO has antibacterial properties, the effect of the nanocomposite, combined with the effect of *S. oneidensis* MR-1 on the reduction of 4-nitrophenol, enabled the reaction to reach 98.2% completion in 10 min (Song and shi 2019).

In another study, graphene oxide reduction was mediated by the leaf extract of *Lantana camara*. The antimicrobial effect of the reduced graphene oxide was evaluated against Gram-positive (*Bacillus subtilis*, *Staphylococcus epidermis*) and Gram-negative (*E. coli*, *Pseudomonas aeruginosa*) bacterial pathogens. An effective antibacterial activity against Gram-positive bacterial pathogens was observed. The authors suggest that the antimicrobial effect was due to the adherence of the material to the surface of the bacterial cell, which led to the rupture of the cell membrane and subsequently to oxidative stress. In addition, phytochemical constituents of *L. camara* extract may have been responsible for increasing the antibacterial effects (Thiyagarajulu and Arumugam 2020).

9.6 CONCLUSIONS

In this chapter, some green approaches for the synthesis of graphene and graphene oxide have been presented. The characteristics and properties of these materials have been discussed. It was shown in this chapter that graphene and graphene oxide materials can be synthesized by different methodologies (mechanical or chemical exfoliation, chemical vapour deposition, epitaxial growth, and chemical methods), where each method produces materials with a particular structure and properties. The choice of the method must consider the necessary amount of material and the desired application. However, traditional synthesis has advantages such as the possibility of large-scale production and of generating high-quality graphene with high crystallinity. It has some drawbacks, such as the need for high temperatures, time consumption, low material homogeneity, substrate effect, problems with large-scale production, and the use of strong oxidizers and concentrated acids. In this chapter, green alternatives have been described to minimize these disadvantages. The green synthesis of graphene mediated by plant extract, proteins, or microorganisms has been discussed in this chapter. From this approach, it was shown that green strategies appear to be potential alternatives to traditional synthesis. Finally, due to the concern regarding the effect of antimicrobial resistance on the health of the world population, this chapter has pointed out the potential of green graphene-based materials as antimicrobial agents.

9.7 REFERENCES

Agarwal, V., and P. B. Zetterlund. 2021. Strategies for reduction of graphene oxide: A comprehensive review. *Chemical Engineering Journal*, no. 405 (February): 127018. https://doi.org/10.1016/j.cej.2020.127018.

Anand, A., B. Unnikrishnan, S. C. Wei, C. P. Chou, L. Z. Zhang, and C. C. Huang. 2019. Graphene oxide and carbon dots as broad-spectrum antimicrobial agents: A minireview. *Nanoscale Horiz*, no. 4 (September): 117–137. https://doi.org/10.1039/c8nh00174j.

Azizi-Lalabadi, M., H. Hashemi, J. Feng, and S. M. Jafari. 2020. Carbon nanomaterials against pathogens; the antimicrobial activity of carbon nanotubes, graphene/graphene oxide, fullerenes, and their nanocomposites. *Advances in Colloid and Interface Science*, no. 284 (October): 102250. https://doi.org/10.1016/j.cis.2020.102250.

Bhardwaj, B., P. Singh, A. Kumar, S. Kumar, and V. Budhwar. 2020. Eco-friendly greener synthesis of nanoparticles. *Advanced Pharmaceutical Bulletin*, no. 10 (August): 566–576. https://doi.org/10.34172/apb.2020.067.

Brodie, B. C. 1859. XIII. On the atomic weight of graphite. *Philosophical Transactions of the Royal Society of London*, no. 149 (January): 249–259. https://doi.org/10.1098/rstl.1859.0013.

Chamoli, P., R. Sharma, M. K. Das, and K. K. Kar. 2016. Mangifera indica, Ficus religiosa and Polyalthia longifolia leaf extract-assisted green synthesis of graphene for transparent highly conductive film. *RSC Advances*, no. 6 (September): 96355–96366. https://doi.org/10.1039/C6RA19111H.

Chen, Y., Y. Niu, T. Tian, J. Zhang, Y. Wang, Y. Li, and L. Qin. 2017. Microbial reduction of graphene oxide by Azotobacter chroococcum. *Chemical Physics Letters*, no. 677 (June): 143–147. https://doi.org/10.1016/j.cplett.2017.04.002.

Cherian, R. S., S. Sandeman, S. Ray, I. N. Savina, J. Ashtami, and V. P. Mohanan 2019. Green synthesis of Pluronic stabilized reduced graphene oxide: Chemical and biological characterization. *Colloids Surf B Biointerfaces*, no. 179 (July): 94–106. https://doi.org/10.1016/j.colsurfb.2019.03.043.

Chettri, P., Vendamani, V. S., Tripathi, A., Pathak, A. P., & Tiwari, A. (2016). Self assembly of functionalised graphene nanostructures by one step reduction of graphene oxide using aqueous extract of Artemisia vulgaris. *Applied Surface Science*, *362*, 221–229.

Choi, W., I. Lahiri, R. Seelaboyina, and Y. S. Kang. 2010. Synthesis of graphene and its applications: A review. *Critical Reviews in Solid State and Materials Sciences*, no. 35 (February): 52–71. https://doi.org/10.1080/10408430903505036.

Cruz, D. M., E. Mostafavi, A. Vernet-Crua, H. Barabadi, V. Shah, J. L. Cholula-Díaz, G. Guisbiers, and T. J. Webster. 2020. Green nanotechnology-based zinc oxide (ZnO) nanomaterials for biomedical applications: A review. *Journal of Physics: Materials*, no. 3 (May): 034005. https://doi.org/10.1088/2515-7639/ab8186.

Gurunathan, S., J. W. Han, V. Eppakayala, and J.-H. Kim. 2013a. Green synthesis of graphene and its cytotoxic effects in human breast cancer cells. *International Journal of Nanomedicine*, no. 8 (March): 1015–1027. https://doi.org/10.2147/IJN.S42047.

Gurunathan, S., J. W. Han, V. Eppakayala, and J. Kim. 2013b. Microbial reduction of graphene oxide by Escherichia coli: A green chemistry approach. *Colloids and Surfaces B: Biointerfaces*, no. 102 (February): 772–777. https://doi.org/10.1016/j.colsurfb.2012.09.011.

Gurunathan, S., J. W. Han, J. H. Park, and J. H. Kim. 2014. An in vitro evaluation of graphene oxide reduced by Ganoderma spp. in human breast cancer cells (MDA-MB-231). *International journal of nanomedicine*, no. 9 (April): 1783–1797.

Hassanien, R., D. Z. Husein, and M. Khamis. 2019. Novel green route to synthesize cadmium oxide@graphene nanocomposite: Optical properties and antimicrobial activity. *Materials Research Express*, no. 6 (May): 085094. https://doi.org/10.1088/2053-1591/ab23ac.

Hou, D., Liu, Q., Wang, X., Quan, Y., Qiao, Z., Yu, L., & Ding, S. (2018). Facile synthesis of graphene via reduction of graphene oxide by artemisinin in ethanol. *Journal of Materiomics*, *4*(3), 256-265.

Hummers, W. S., and R. E. Offeman. 1958. Preparation of graphitic oxide. *Journal of the American Chemical Society*, no. 80 (March): 1339. https://doi.org/10.1021/ja01539a017.

Ismail, Z. 2019. Green reduction of graphene oxide by plant extracts: A short review. *Ceramics International*, no. 45 (December): 23857–23868. https://doi.org/10.1016/j.ceramint.2019.08.114.

Keerthi, M., M. Akilarasan, S.-M. Chen, S. Kogularasu, M. Govindasamy, V. Mani, M. A. Ali, F. M. A. Al-Hemaid, and M. S. Elshikh. 2018. One-pot biosynthesis of reduced graphene oxide/prussian blue microcubes composite and its sensitive detection of prophylactic drug dimetridazole. *Journal of The Electrochemical Society*, no. 165 (January): B27–B33. https://doi.org/10.1149/2.0591802jes.

Khanra, P., T. Kuila, N. H. Kim, S. H. Bae, D. Yu, and J. H. Lee. 2012. Simultaneous bio-functionalization and reduction of graphene oxide by baker's yeast. *Chemical Engineering Journal*, no. 183 (February): 526–533. https://doi.org/10.1016/j.cej.2011.12.075.

Khojasteh, H., Safajou, H., Mortazavi-Derazkola, S., Salavati-Niasari, M., Heydaryan, K., & Yazdani, M. (2019). Economic procedure for facile and eco-friendly reduction of graphene oxide by plant extracts; a comparison and property investigation. *Journal of Cleaner Production*, *229*, 1139–1147.

Kim, E., X. Qin, J. B. Qiao, Q. Zeng, J. D. Fortner, and F. Zhang. 2020. Graphene oxide/mussel foot protein composites for high-strength and ultra-tough thin films. *Scientific Reports*, no. 10 (November): 19082. https://doi.org/10.1038/s41598-020-76004-6.

Lee, X. J., B. Y. Z. Hiew, K. C. Lai, L. Y. Lee, S. Gan, S. Thangalazhy-Gopakumar, and S. Rigby. 2019. Review on graphene and its derivatives: Synthesis methods and potential industrial implementation. *Journal of the Taiwan Institute of Chemical Engineers*, no. 98 (May): 163–180. https://doi.org/10.1016/j.jtice.2018.10.028.

Lim, J. Y., N. M. Mubarak, E. C. Abdullah, S. Nizamuddin, M. Khalid, and Inamuddin. 2018. Recent trends in the synthesis of graphene and graphene oxide based nanomaterials for removal of heavy metals: A review. *Journal of Industrial and Engineering Chemistry*, no. 66 (October): 29–44. https://doi.org/10.1016/j.jiec.2018.05.028

Lingaraju, K., H. R. Naika, G. Nagaraju, and H. Nagabhushana. 2019. Biocompatible synthesis of reduced graphene oxide from Euphorbia heterophylla (L.) and their in-vitro cytotoxicity against human cancer cell lines. *Biotechnology Reports*, no. 24 (December): e00376. https://doi.org/10.1016/j.btre.2019.e00376.

Liu, R., Qin, P., Wang, L., Zhao, X., Liu, Y., & Hao, X. (2010). Toxic effects of ethanol on bovine serum albumin. *Journal of Biochemical and Molecular Toxicology*, *24*(1), 66–71.

Maddinedi, S. B., Mandal, B. K., Vankayala, R., Kalluru, P., Tammina, S. K., & Kumar, H. K. (2014). Casein mediated green synthesis and decoration of reduced graphene oxide. *Spectrochimica Acta Part A: Molecular and Biomolecular Spectroscopy*, *126*, 227–231.

Mhamane, D., W. Ramadan, M. Fawzy, A. Rana, M. Dubey, C. Rode, B. Lefez, B. Hannoyer, and S. Ogale. 2011. From graphite oxide to highly water dispersible functionalized graphene by single step plant extract-induced deoxygenation. *Green Chemistry*, no. 13 (July): 1990–1996. https://doi.org/10.1039/C1GC15393E.

Nichols, F., and S. Chen. 2020. Graphene oxide quantum dot-based functional nanomaterials for effective antimicrobial applications. *The Chemical Record*, no. 20 (September): 1505–1515. https://doi.org/10.1002/tcr.202000090.

Novoselov, K. S., A. K. Geim, S. V. Morozov, D. Jiang, Y. Zhang, S. V. Dubonos, I. V. Grigorieva, and A. A. Firsov. 2004. Electric field effect in atomically thin carbon films. *Science*, no. 306 (October): 666–669. https://doi.org/10.1126/science.1102896

Ousaleh, H. A., Charti, I., Sair, S., Mansouri, S., Abboud, Y., & El Bouari, A. (2020). Green and low-cost approach for graphene oxide reduction using natural plant extracts. *Materials Today: Proceedings*, *30*, 803–808.

Rakhimol, K. R., S. Thomas, N. Kalarikkal, and K. Jayachandran. 2020. Casein mediated synthesis of stabilized metal/metal-oxide nanoparticles with varied surface morphology through pH alteration. *Materials Chemistry and Physics*, no. 246 (May): 122803. https://doi.org/10.1016/j.matchemphys.2020.122803.

Rana, A., K. Yadav, and S. Jagadevan. 2020. A comprehensive review on green synthesis of nature-inspired metal nanoparticles: Mechanism, application and toxicity. *Journal of Cleaner Production*, no. 272 (November): 122880. https://doi.org/10.1016/j.jclepro.2020.122880.

Rojas-Andrade, M. D., T. A. Nguyen, W. P. Mistler, J. Armas, J. E. Lu, G. Roseman, W. R. Hollingsworth, F. Nichols, G. L. Millhauser, A. Ayzner, C. Saltikov, and S. Chen. 2020. Antimicrobial activity of graphene oxide quantum dots: impacts of chemical reduction. *Nanoscale Advances*, no. 2 (January):1074–1083. https://doi.org/10.1039/c9na00698b.

Sadhasivam, S., V. Vinayagam, and M. Balasubramaniyan. 2020. Recent advancement in biogenic synthesis of iron nanoparticles. *Journal of Molecular Structure*, no. 1217 (October): 128372. https://doi.org/10.1016/j.molstruc.2020.128372.

Salas, E. C., Z. Sun, A. Lüttge, and J. M. Tour. 2010. Reduction of graphene oxide via bacterial respiration. *ACS Nano*, no. 4 (July): 4852–4856. https://doi.org/10.1021/nn101081t.

Schedin, F., A. K. Geim, S. V. Morozov, E. W. Hill, P. Blake, M. I. Katsnelson, and K. S. Novoselov. 2007. Detection of individual gas molecules adsorbed on graphene. *Nature Materials*, no. 6 (July): 652–655. https://doi.org/10.1038/nmat1967.

Shah, R., A. Kausar, B. Muhammad, and S. Shah. 2015. Progression from graphene and graphene oxide to high performance polymer-based nanocomposite: A review. *Polymer-Plastics Technology and Engineering*, no. 54 (September): 173–183. https://doi.org/10.1080/03602559.2014.955202.

Somani, P. R., S. P. Somani, and M. Umeno. 2006. Planer nano-graphenes from camphor by CVD. *Chemical Physics Letters*, no. 430 (October): 56–59. https://doi.org/10.1016/j.cplett.2006.06.081.

Somu, P., and S. Paul. 2019. A biomolecule-assisted one-pot synthesis of zinc oxide nanoparticles and its bioconjugate with curcumin for potential multifaceted therapeutic applications. *New Journal of Chemistry*, no. 43 (June): 11934–11948. https://doi.org/10.1039/c9nj02501d.

Song, T., Xu, H., Wei, C., Jiang, T., Qin, S., Zhang, W., ... & Cao, Y. (2016). Horizontal transfer of a novel soil agarase gene from marine bacteria to soil bacteria via human microbiota. *Scientific Reports*, *6*(1), 1–10.

Song, X., and X. Shi. 2019. Biosynthesis of Ag/reduced graphene oxide nanocomposites using Shewanella oneidensis MR-1 and their antibacterial and catalytic applications. *Applied Surface Science*, no. 491 (October): 682–689. https://doi.org/10.1016/j.apsusc.2019.06.154.

Staudenmaier, L. 1898. Verfahren zur Darstellung der Graphitsäure. *Berichte der deutschen chemischen Gesellschaft*, no. 31 (October): 1481–1487. https://doi.org/10.1002/cber.18980310237.

Thakur, S., and N. Karak. 2015. Alternative methods and nature-based reagents for the reduction of graphene oxide: A review. *Carbon*, no. 94 (November): 224–242. https://doi.org/10.1016/j.carbon.2015.06.030.

Thiyagarajulu, N., and S. Arumugam. 2020. Green synthesis of reduced graphene oxide nanosheets using leaf extract of lantana camara and its in-vitro biological activities. *Journal of Cluster Science*, (May): 1–10. https://doi.org/10.1007/s10876-020-01814-7

Vargas, C., R. Simarro, J. A. Reina, L. F. Bautista, M. C. Molina, and N. González-Benítez. 2019. New approach for biological synthesis of reduced graphene oxide. *Biochemical Engineering Journal*, no. 151 (November): 107331. https://doi.org/10.1016/j.bej.2019.107331.

Viculis, L. M., J. J. Mack, and R. B. Kaner. 2003. A chemical route to carbon nanoscrolls. *Science*, no. 299 (February): 1361. https://doi.org/10.1126/science.1078842.

Wang, H., W. Huang, S. Huang, L. Xia, Xinyue Liu, Y. Li, S. Song, and L. Yang. 2021. A green method toward graphene oxide reduction by extracellular polymeric substances assisted with NH_4^+. *Arabian Journal for Science and Engineering*, no. 46 (September): 485–494. https://doi.org/10.1007/s13369-020-04936-2.

Weng, X., Z. Lin, X. Xiao, C. Li, and Z. Chen. 2018. One-step biosynthesis of hybrid reduced graphene oxide/iron-based nanoparticles by eucalyptus extract and its removal of dye. *Journal of Cleaner Production*, no. 203 (December): 22–29. https://doi.org/10.1016/j.jclepro.2018.08.158.

Zhu, Y., S. Murali, W. Cai, X. Li, J. W. Suk, J. R. Potts, and R. S. Ruoff. 2010. Graphene and graphene oxide: Synthesis, properties, and applications. *Advanced Materials*, no. 22 (June): 3906–3924. https://doi.org/10.1002/adma.201001068.

10 A Review on Fullerenes and its Applications in Health Care Sector

M. Sundararajan, L. Athira, R. A. Renjith, M. Prasanna, R. G. Rejith, S. Ramaswamy, Sarika Verma, and M. A. Mohammed-Aslam

CONTENTS

10.1 INTRODUCTION

Fullerene is an allotrope of carbon. These carbon atoms are connected with single and double bonds to form a partially closed loop. Fullerene contains five- to seven-membered fused rings. It is composed of sheets of hexagonal rings which prevent the structure from being planar. The system of fullerene molecules may be ellipsoid,

DOI: 10.1201/9781003110781-10

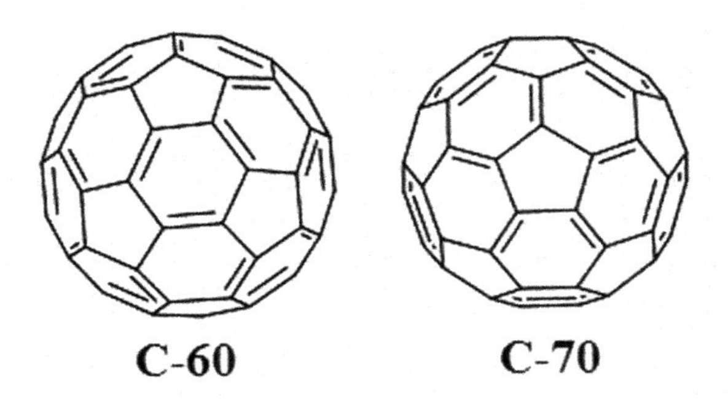

FIGURE 10.1 Structure of C_{60} and C_{70}.

hollow spherical, cylindrical, or any other sizes and shapes. The spherical fullerene is also called Buckminsterfullerene or buckyballs. The molecule was named after the architect Richard Buckminster Fuller, who created the geodesic dome. Buckminsterfullerene has a similar structure to that of a dome. Cylindrical fullerene is called carbon nanotubes or bucky tubes. The empirical formula of buckminsterfullerene is C_{60}, as it contains 60 carbon atoms. It has a truncated icosahedron structure. It contains 20 hexagons and 12 pentagons. Another fullerene (C_{70}), which contains 70 carbon atoms, is also present, and is similar in shape to a rugby ball. It comprises 25 hexagons and 12 pentagons (Qiao et al. 2007). The structures of C_{60} and C_{70} are given in Figure 10.1.

Fullerene has structural similarity with graphite, in which each carbon atom is covalently bonded to three of its neighbours. Hence, it conducts electricity. Fullerene has outstanding mechanical rigidity. The fullerite crystal is very soft under normal conditions, but, under pressure, it changes to a hard, rigid material harder than diamond. This occurs due to 3-D polymerization. Fullerenes are extensively used for biomedical applications such as drug delivery, X-ray imaging, high-performance MRI, and so on (Lalwani and Sitharaman 2013). Fullerenes are widely used for cancer treatment. Fullerene can be absorbed by the tumour cells with the help of functional groups such as folic acid, L-arginine, and L-phenylalanine. Once it is absorbed and exposed to light radiation, it eliminates the DNA, lipids, and proteins that make up the cancer cell (Brown et al. 2004). Fullerenes have been a subject in material science, nanotechnology, and electronics of intense research for both their technological aspects and their chemistry (Belkin et al. 2015).

During the past decade, environmental contamination has increased significantly. This has enhanced interest in the engineering non-materials such as fullerene C_{60} due to its physical and chemical properties. Partly, they can induce mutagenesis. Carbon-based nanomaterials have a wide range of applications, such as drug delivery, polymer additive, fuel cells, and cosmetics (Chen et al. 2008). A small number of analytical techniques are used for the monitoring of fullerenes in the terrestrial and aquatic geological area. The most widely used method for detecting fullerene is high-performance liquid chromatography (HPLC) (Shareef et al. 2010).

Fullerene was first reported from geological materials in Karelian shungite, a carbon-rich rock (Buseck 2002, Mossman et al. 2003, Mosin and Ignatov 2013).

Fullerene content in geological matter ranges from approximately μg kg^{-1} to mg kg^{-1}. The results obtained from various research studies were inconsistent, with low reproducibility. For example, C_{60} was detected from the Sudbury impact structure at about 10 mg kg^{-1} (Becker et al. 1994). The same was detected from same source in a second sample, but at much lower concentrations, about 2 μg kg^{-1} (Elsila et al. 2005). But in the third sample, no fullerene was detected (Heymann et al. 1999), even though researchers had been trying to extract and identify fullerene from geological materials.

After the discovery of natural fullerene, C_{60} was identified from several geological materials, including Karelia, Russia, the Bohemian Massif, Ontario in Canada, Sheep Mountain, Colorado in the USA, The Cretaceous-Tertiary boundary, and Permian-Triassic boundary. The analytical methods used for the characterization of fullerene were HPLC, mass spectroscopy, HR-TEM (High Resolution Transmission Electron Microscopy), and Raman spectroscopy. This paper reviews the different places where natural fullerene has been identified and the methods through which it was extracted.

10.2 FULLERENE FORMATION THEORIES

Many models for natural fullerene formation have been proposed, although, due to the absence of mass occurrence of fullerene and difficulties in testing the hypotheses, questions remain. Among the proposals, concepts such as lightning strikes or stellar interiors were discussed by early workers. Many constraints affect fullerene formation, such as inhibitory factors like oxygen, nitrogen, and other non-inert gases. The Precambrian coal-rich shungite, which is the primary source of fullerene, is strongly affected by metamorphism (Volkova and Bogdanova 1986). The meteorite impact model of fullerene formation was postulated by Becker and Luann (1994) after a thorough analysis of the Sudbury impact basin. The enigmatic model of fullerene occurrence in space is discussed by Zhang et al. (2020). The noble gas carrier model of fullerene in the meteorite is discussed by Poreda and Becker (2003), and the presence of C_{60} to C_{400} in carbonaceous chondrite meteorites supports this model. Various forms of fullerene and carbon nanorods have been isolated from asphaltenes from crude oil samples (Santos et al. 2016), whereas the organic model of fullerene synthesis through coal formation have been well studied in Yunnan Province, China (Fang and Wong 1997).

10.2.1 Pyrometamorphism Model of C_{60}

This model was first discussed in a detailed study of Coorongite, Australia. Scientists postulated the high (900–1200 °C) transformation temperatures of carbon precursors (the most common one is bitumen and petroleum asphaltenes). Alga-derived bitumen with oxygen-poor kerogen precursor is also an example of this. Some of the Archean-Neoproterozoic black shale deposits gave evidence supporting this model, such as Cretaceous Rundle shale (Australia), Guayas Basin, where high temperature (700–900 °C) pyrolysis of biogenic in black shale along with basaltic extrusions is observed. The Absence of C_{70} evident the solid transformation of carbon precursors.

10.2.2 Biogenic Formation Models

Heymann et al. (2003) proposed the biogenic model of fullerene formation. They postulated that polycyclic aromatic hydrocarbons (PAHs), which are formed from algae, were transformed to fullerene during regional metamorphism. Later, many researchers produced synthetic fullerene from organic sources such as the rubbery algal product coorongite. But the synthetic pathways of natural products have not been well studied.

10.3 SITES OF NATURALLY OCCURRING FULLERENE

Fullerene occurred naturally in (i) Sheep Mountain, Colorado, (ii) the Cretaceous-Tertiary boundary sediments, (iii) Decan-Anjar, Kutch, India, (iv) Yunnan, southern China, (v) Kondopoga, Karelia, Russia, (vi) the Permo-Triassic (P/T) boundary-Inuyama, Central Japan, (vii) the Bohemian Massif, Czech Republic, (viii) the Ries Impact Crater, Bavaria, Germany, (ix) the Mangampeta Baryte Mine, Andhra Pradesh, India, (x) Llobregat River, Spain, (xi) Riyadh, Yanbu' Al Bahr, Jubail, Al Diriyah in Saudi Arabia, (xii) soil collections from Amsterdam, the Netherlands, and (xiii) the Sava River, southern Europe. Table 10.1 lists sites of naturally occurring fullerene.

10.3.1 Sheep Mountain, Colorado, USA

Fulgurites are glassy, rock-like materials that contain fullerene. Basically, Sheep Mountain was formed by Precambrian crystalline rocks consisting mainly of mafic and felsic gneisses, and schist that has been metamorphosed into amphibolite facies, metasedimentary, and metavolcanic which was intruded by Sherman Granite. These are formed mainly by lightning. When lightning strikes the ground, it produces conditions similar to those used in the laboratory to synthesize fullerenes. The fulgurites found in Sheep Mountain are formed by lightning, and the resulting form has a dendritic structure. The sample to be examined was pulverized, followed by extraction in the Soxhlet apparatus for 24 hours. After each extraction, all the glassware was appropriately cleaned and baked. Then the solvent was evaporated, and the extract was dissolved again in 1 ml of the solvent, and taken for analysis. The fulgurites from the sheep mountain contained a detectable amount of fullerene. Non-fused sample from the host rock was subjected to solvent extraction, but fullerene was not detected. It shows that contamination or instrumental error were not the sources of the fullerene detected. The exact parameter of lightning which contributes to the formation of fullerene is not known accurately, but an average consideration could be estimated. The damaging cloud-to-ground strike is the reason for 90% of all lightning which hits the ground. Hence, a tremendous amount of energy is produced for the reaction. The cloud-to-ground lightning travels as a stepladder . The potential

TABLE 10.1
Naturally Occurring Fullerene Sites

Sl. No.	Location	Country	Analytical Technique	Inference	References
1	Sheep Mountain	USA	Mass spectrosopy	m/z =720 amu for C_{60} m/z =840 amu for C_{70}	Daly et al. (1993)
2	Cretaceous-Tertiary boundary sediments		—	—	Heymann et al. (1994a, b), (1996), Heymann and Wolbach (2006)
2.1	The Sumbar	Turkmenistan	HPLC, mass spectroscopy	m/z =720 amu for C_{60} m/z =840 amu for C_{70} C_{70}/C_{60} = 0.21–0.23	—
2.2	The Malyi Balkhan	Turkmenistan	HPLC, mass spectroscopy	m/z =720 amu for C_{60} m/z =840 amu for C_{70} C_{70}/C_{60} = 0.36	—
2.3	Koshak	Kazakhstan	HPLC, mass spectroscopy	m/z =720 amu for C_{60}	—
2.4	Tetra Tskaro	Georgia	HPLC, mass spectroscopy	m/z =720 amu for C_{60}	—
2.6	The sea cliff of Stevns Klint	Austria	HPLC, mass spectroscopy	m/z =720 amu for C_{60} m/z =840 amu for C_{70} C_{70}/C_{60} = 0.30–0.33	—

(Continued)

TABLE 10.1 (CONTINUED)
Naturally Occurring Fullerene Sites

Sl. No.	Location	Country	Analytical Technique	Inference	References
2.7	Elendgraben	Denmark	HPLC, mass spectroscopy	m/z =720 amu for C_{60}	—
2.8	The Caravaca	Spain	HPLC, mass spectroscopy	m/z =720 amu for C_{60} m/z =840 amu for C_{70} C_{70}/C_{60} = 0.22–0.26	—
2.9	The Flaxbourne River	New Zealand	HPLC, mass spectroscopy	m/z =720 amu for C_{60} m/z =840 amu for C_{70} C_{70}/C_{60} = 0.28–0.36	—
3	Decan-Anjar, Kutch	India	FTIR, ^{13}C- NMR spectroscopy, mass spectroscopy	m/z =720 amu for C_{60}	Ghevariya and Sundaram (1996), Bhandari et al. (2002)
4	Yunnan	Southern China	HPLC	C_{60}:C_{70} =3.2	Fang and Wong (1997), Fang et al. (2006)
5	Kondopoga, Karelia	Russia	^{1}H NMR spectroscopy, mass spectroscopy	Single line peak at 143.2 corresponds to C_{60} in NMR and m/z = 720 amu	Parthasarathy et al. (1998)
6	Permo-Triassic (P/T) boundary-Inuyama	Central Japan	HPLC	Peak at 332 nm on chromatogram	Chijiwa et al. (1999)

(Continued)

TABLE 10.1 (CONTINUED)
Naturally Occurring Fullerene Sites

Sl. No.	Location	Country	Analytical Technique	Inference	References
7	Bohemian Massif	Czech Republic	HPLC, mass spectroscopy, Raman spectra	*m/z* =720 amu and bands at 1350, 1575, 2680, and 2930 cm^{-1} corresponds to carbonaceous matter	Jehlička et al. (2003)
8	Ries impact crater, Bavaria	Germany	HPLC, MALDI-TOF	No peak at *m/z* =720 amu, absence of C_{60}	Frank et al. (2005)
9	Mangampeta Baryte Mine, Andhra Pradesh	India	Mass spectrosopy	*m/z* =720 amu for C_{60} *m/z* =840 amu for C_{70}	Misra et al. (2007)
10	Llobregat River	Spain	HPLC, triple quadrupole-mass spectroscopy	*m/z* =720 amu for C_{60} *m/z* =840 amu for C_{70}	Sanchís et al. (2013)
11	Urban and industrial area	Saudi Arabia	HPLC, triple quadrupole-mass spectroscopy	*m/z* = 720 amu for C_{60}	Sanchís et al. (2013)
12	Amsterdam	The Netherlands	Mass spectroscopy	*m/z* = 720 amu for C_{60}	Carboni et al. (2016)
13	Sava River	Southern Europe	Mass spectroscopy	*m/z* = 720 amu for C_{60} *m/z* =840 amu for C_{70}	Barceló Cullerés (2018)

difference between the ground and the leader may exceed 10 million volts. The vapour atmosphere in which fullerene formation occurs is essential. The first step is carbon vapourization. In this sample, the gas environment was a type different from that present at the time of the lightning (Daly et al. 1993).

10.3.2 Cretaceous-Tertiary Boundary Sediments

Samples collected from various sites at the Cretaceous-Tertiary boundary (KTB) were crushed. Then, the carbonates from the samples were removed by treatment with 1: 9 ratio of concentrated HCl: water until the evolution of carbon dioxide ceased. The obtained suspension was filtered, and the residue washed with 1:1 37% HCl and 48% HF and then dried. Silicates present were also removed with 1:1 37% HCl: 48% HF. All glassware was sonicated in toluene except the injection syringe, before being dried and heated to 350 °C for at least 24 hours. The demineralized samples were analyzed by HPLC. The extracted sample was approximately 0.3–15 ppb. The maximum amount was from the Flaxbourne River (Heymann et al. 1994a, b, 1996, 1999, 2003, 2006). It was proposed that wildfires formed the fullerenes found at the KTB following extraction. Fullerene was deposited, along with the soot particles, on the surface to which they were adsorbed. It was also suggested that the fullerenes were formed from confined fires, rather than from worldwide deposits. It was not concluded why fullerenes were not formed at specific locations. The most logical explanation was that the local conditions were unfavourable for fullerene-producing wildfires.

10.3.3 Decan-Anjar, Kutch, India

Toluene-soluble and toluene-insoluble fullerene C_{60} were extracted from iridium-rich sediments at Anjar, India, geologically located in the western margin of the Deccan lava province. The section contains nine lava flows and a minimum of four Intertrappean Beds. Mainly consisting of limestone, shale, and mudstone, the Limnotic layer reports higher iridium concentrate to correlate with the Cretaceous-Tertiary (KT) boundary. The Cretaceous-Tertiary boundary is found at the third and fourth lava flow (Ghevariya and Sundaram 1996). Samples collected from the third Inter trappean Bed were characterized with the help of mass spectroscopy (Bhandari et al. 2002).

10.3.4 Yunnan, Southern China

Fullerene was extracted from coal in Yunnan Province, southern China. Coal seams of Yunnan province largely come under the depression of the Precambrian age to the Tertiary basin, an area lithologically dominated by Palaeozoic sedimentary rocks. The age of the coal was predominantly from the upper Miocene to the lower Pliocene. From several coal mining regions, two types of coal were collected, which were designated as *B* and *K*. *B* consisted of hard granules and was bright. The carbon content was high in *B*, but the extracted fullerene was greater in *K*. In both *B* and *K*, the C_{60} percentage was greater than the C_{70} percentage. In *K*, 74% C_{60} and, in *B*,

85% C_{60} were extracted. The HPLC analysis carried out for both *B* and *K* in the collected samples showed a prominent C_{60} peak in both samples, but not for C_{70}. The collected concentration of fullerene of the *K* coal type was 2.62×10^{-4} ppm, and for *B* was 2.64×10^{-5} ppm for 65 g of *K* and 151.1 g of *B* (Fang and Wong 1997; Fang et al. 2006).

10.3.5 Kondopoga, Karelia, Russia

The Proterozoic sequence of Karelia is famous for its deposits of carbon-rich rocks called shungite. They have been exploited for the manufacture of lightweight aggregates. The Proterozoic rocks found in Karelia, mainly the lower part, are divided into Sariolian, Sumain, Ludicovian, Jjatulian Livvian, and Vespian subgroups. The shungite deposits are present in Ludicovian and Livvian subgroups. Using a standard acid dissolution method, the carbonaceous matter was extracted from the shungite sample. The residue thus obtained was washed with distilled water. The residue was analyzed by ^{13}C-NMR, mass spectroscopy, and powder X-ray diffraction. The mass spectrum of the sample could not detect any C_{70} content. The spectrum obtained was centroid. A single scan can provide enough signal of C_{60} fullerene. Using ^{13}C-NMR, the presence of fullerene in the shungite sample was established (Parthasarathy et al. 1998).

10.3.6 Permo-Triassic (P/T) boundary-Inuyama, Central Japan

The rock samples were collected from early Triassic chert and claystone above the P/T horizon from the Inuyama area. Panthalassan pelagic deep seafloor deposits consist of bedded chert, thin, tabular shale, jet black shale, grey siliceous shale, chert, and breccia. The samples extracted with solvent were crushed, evaporated, and filtered. The collected rock samples were crushed to less than 0.5 mm in diameter. Samples were treated with HF or HCl solution, which is 50% aqueous. Then, the carbon residue containing any fullerene present was extracted with toluene using an ultrasonic bath at room temperature for 5–10 hours, followed by concentration by rotary evaporation down to 1 ml or less. HPLC analysis was then carried out at room temperature, where C_{60} was detected only in P/T boundary samples. The presence of combustion at the boundary period is considered to be the main reason for the occurrence of fullerene. The most considerable combustion was caused by wildfire, which are necessary to provide the harsh extreme conditions required for the formation of natural fullerene. The extracted fullerene had a C_{60} content greater than the C_{70} content. The complete HPLC identification, time of retention, and UV-Visible absorption spectra confirmed the presence of fullerene in the collected samples (Chijiwa et al. 1999).

10.3.7 Bohemian Massif, Czech Republic

Highly carbonized, glass-like, solid bitumen was reported from the sedimentary volcano sequence of the Neoproterozoic age of the Bohemian Massif. Solid bitumen was reported in pillow lava, which is the central part of a volcano-sedimentary sequence of the Neoproterozoic age. About 3 g of the collected sample was pulverized in an agate motor by grinding. It was then immersed in the solvent carbon disulphide (CS_2) and then treated in an ultrasonic bath for 1 hour with vigorous shaking at room temperature in an Ar atmosphere. The extracted samples were used for HPLC analysis. All the agate motors should be new and cleaned by heating at 500 °C for 12 hours to avoid contamination. For HPLC analysis, the selected column should have dimensions of 25 cm length and a 2.1 mm inner diameter. The column was then packed with octadecyl silica, with a 1:1 toluene-methanol mixture being used as the mobile phase. The volume of the sample to be injected onto the column was 20 μL. From 3 g of the sample, 0.5 ml of the final extract, and 20 μL loading volume, about 0.01 ppm of C_{60} were obtained. To prevent contamination of the geological sample with synthetic reference fullerene, the following precautions had to be carried out: (i) one blank should be run before each experiment, (ii) the fullerene content in the extract of the geological sample should be measured twice, and (iii) the sequence of the analyses should be terminated by running a blank. This sequence was repeated for each sample (Jehlick et al. 2003).

10.3.8 Ries Impact Crater, Bavaria, Germany

The Ries Crater (Figure 10.2) is in southern Germany. The middle Miocene crater was formed by an achondrite type of bolide around 15 million years ago, and showed the presence of impact origin minerals like coesite and suevite. The crater has a circular shape with an inner diameter of 25 km. Samples from the Ries Crater were collected. The samples taken for analysis were fresh and non-weathered. The

FIGURE 10.2 . Ries Crater (48°51'06"N, 10°29'23"E) lies in southern Germany.

whole rock sample was used for analysis. The rock sample was powdered, and it was analyzed for TC (total carbon) and organic carbon (OC) content. Further analysis was done by HPLC. The process involved three main steps: demineralization, extraction, and analysis. Using concentrated hydrochloric acid and hydrofluoric acid, demineralization was carried out. The extraction process was performed with HPLC grade toluene. The sample was treated for 2 hours in an ultrasonic bath. Then, the solution was filtered and evaporated under vacuum to 1 mL. HPLC was used for analysis, in which toluene is used isocratically as the mobile phase. Fullerene-spiked samples were prepared to quantify any losses incurred during the demineralization and extraction processes (Frank et al. 2005).

10.3.9 Mangampeta Baryte Mine, Andhra Pradesh, India

The black carbonaceous slates in the shungite suite of rocks at Mangampeta in Andhra Pradesh, India (Figure 10.3) contain fullerene. Baryte occurrence is associated with Vempalle dolomite and dolomitic limestones, quartzite Volcanogenic grey granular, lapilli, vein, and replacement, which are important potential sources of fullerene. The baryte deposit of Mangampeta consists of interlayers with thick bands of black slate. Further studies identified that these carbonaceous slates are fullerene bearing and grouped in the shungite suite of rocks. The occurrence of fullerene in the Mangampeta was identified using laser desorption/ionization spectrometry. The carbon-hydrogen-sulphur composition of the acid-digested carbonaceous residue showed that 10–15 wt% was carbon, 1.95 to 2.6 wt% was sulphur, 0.06 to 0.08 wt% was H/C ratio, and 5 to 9.25 wt% was C/S ratio. This result indicated that the rock belongs to the shungite suite of rocks. The laser desorption/ionization mass spectra

FIGURE 10.3 Mangampeta Baryte mine in Andhra Pradesh, India.

of both rock powder and Soxhlet extract, using a toluene extract, showed the presence of fullerene. The *m/z* peaks at 720 amu and 840 amu corresponded to C_{60} and C_{70}. This is the first place in India to be identified with the occurrence of fullerene-bearing shungite suite rock (Misra et al. 2007).

10.3.10 Llobregat River, Spain

River sediments from various sites of the Llobregat River Basin were collected, and water samples from the collection sites were also taken. The extraction technique was ultrasound-assisted liquid extraction. In the soil collected, about 33% C_{60} was detected, whereas, in the collected river water, C_{70} was seen, at about 67%. The samples were collected from the urban area of Catalonia and industrialized sites near wastewater treatment plants (WWTP) (Sanchis et al. 2013).

10.3.11 Riyadh, Yanbu' Al-Bahr, Jubail, Ad Diriya, Saudi Arabia

Some 58 soil samples from various industrial and urban areas of Riyadh, Yanbu' Al-Bahr, Jubail, and Ad Diriyah in Saudi Arabia were collected. An ultrasound technique was used for the extraction, and a liquid extraction method was also used. Toluene was selected as solvent for the extraction technique. In about 19% of the collected samples, only C_{60} was detected. The soil was collected from the sites having high influence from petroleum refineries and from urban sites. The soil samples were dried and placed in a desiccator overnight. The samples were then finely powdered in an agate motor and sieved. Ultrasound extraction was carried out and followed by centrifugation. The toluene extractant was collected, and was concentrated to 5 mL. HPLC was used for further analysis, and mass spectroscopy was also used to confirm the presence of fullerene. Molecular ion value *m/z* = 720 amu confirmed that only fullerene C_{60} was identified in the samples, with the main reason for this being the combustion process (Sanchis et al. 2013).

10.3.12 Soil collections from Amsterdam, the Netherlands

Soil collections from the Netherlands also reported the presence of fullerene. Some 15 samples were collected from an urban area near the coal power plant located in Amsterdam, 16 samples were collected from the AVR incinerator in Duiven, 26 samples were collected from Amsterdam's ring motorway, 16 samples were collected from green areas, such as parks, flower beds etc., of Amsterdam, six samples from the Natural Park at Castricum, and 12 samples from the vicinity of the runway at Eindhoven Airport. All the collected samples were analyzed by HPLC. Biphenyl was used as the stationary phase, and high-purity nitrogen was used as the collision gas for the spectrometer. The mass quadrupole time-of-flight spectrometer and chromatography helped to identify and extract C_{60} fullerene. The main criteria by which fullerene was identified were the chromatographic retention time and the threshold accuracy of mass (Carboni et al. 2016).

10.3.13 Sava River, Southern Europe

The Sava River is a tributary of the Danube River, which is located in southern Europe (Barcelo 2018). Twenty-seven surface water samples and 12 sediment samples were collected from the spot where two extreme hydrologic conditions occurred. About 150 mL of surface water were successively filtered by glass fibre and nylon membrane filters. The filtrate and filters were extracted separately. About 1.5 g of NaCl was used to salt out the filtrate and homogenize it. Then, it was extracted with 50 mL toluene in a separating funnel by liquid-liquid extraction. The filtrate was then dried overnight at 60 °C, followed by extraction of carbonaceous matter in an ultrasonic bath. The sediment samples were extracted with a slight modification of the aforementioned method. The sediments collected were defrosted at room temperature and were wet-sieved. The sieved sediments were then dewatered and centrifuged and then dried in a desiccator after the supernatant was decanted. The dried sample (about 4-5gm) was then made into a homogenized suspension with 50% toluene and allowed to rest in the dark for 3 hours at 4 °C. Then, 40 mL of toluene were added and mixed, and the extraction process was carried out in an ultrasonic bath. HPLC analysis of the extract confirmed the presence of C_{60} fullerene. The presence of fullerene in freshwater as well as in sediment samples was reported, with C_{60} fullerene concentrations of 8 pg/L–59 ng/L and 108–895 pg/g dry weight in water and sediments reported, respectively. Various characterization methods are used to confirm the presence of fullerene. High-resolution transmission electron microscopy (HRTEM), Raman spectroscopy, and mass spectroscopy were mainly used for this purpose. HRTEM was used to obtain information about the dispersion states of fullerene in the geological matrix. C_{60} has hexagonal close-packed diffraction patterns. If any other impurities are present in the sample, the structure will crystallize (Buseck et al. 1992). Based on group theory, it was found that fullerene C_{60} molecules had only ten frequencies that are Raman active, due to the icosahedral symmetry of fullerene. The most intensive band appeared at 1459–1469 cm^{-1}, which is tangential breathing mode Ag. The presence of oxygen atoms in the fullerites may affect the precision of these bands (Buseck et al. 1992).

10.4 APPLICATIONS OF FULLERENES IN HEALTH CARE

The direct application of fullerene and its derivatives indicates promising potential in the field of medicine due to the unusual physical and chemical properties of the fullerene core. The radical sponge character and hydrophobic spheroid of fullerenes are the main cause for their activity in different fields. The low toxicity of fullerene is sufficient to stimulate researchers in biology and in chemistry to integrate their efforts and systematically investigate the biological properties of these fascinating molecules. Research and development activities across the globe have led to large numbers of application-orientated patents, spanning a very broad range of potential commercial applications, including: anticancer drug delivery systems using HIV drugs, photodynamic therapy, and cosmetics to slow down the aging of human skin (Friedman et al. 1993, Sijbesma et al. 1993).

Fullerene C_{60} and its derivatives include all known categories of compounds, explaining its great versatility and higher chemical reaction activities. Derivatives of fullerenes and fullerene C_{60} have the potential to exhibit antiviral activity. They have strong indications of delaying or preventing the onset of AIDS (acquired immune deficiency syndrome). Antiviral compounds of fullerene can suppress the replication of HIV. Hence, it has strong impact on the rehabilitation of patients with HIV. Fullerene and its derivatives possess antioxidant, antiviral, and several other biological properties, on the basis of their unique molecular structure. They are reported to be effective inhibitors of the HIV aspartic protease enzyme and have the ability to form an anti-HIV drug (An et al. 1996).

One another potential application of fullerenes in the medical field is associated with the photoexcitation of fullerenes. In response to photoirradiation, fullerene can be excited from ground level to C_{60}. The life-time of the species is increased by intersystem crossing. $^1C_{60}$ and $^3C_{60}$ (i.e., the higher energy species) are excellent acceptors and, by the process of electron transfer, it can be easily reduced to C_{60}^- in the presence of an electron donor. Under biological conditions, fullerenes can be reduced by reducing agents like guanosine. This property renders them a potential photosensitizer, mainly used in PDT (photodynamic therapy). Also, by using different functional groups, fullerene derivatives have been investigated for potential anticancer activity (Krusic et al. 1991).

Fullerenes are also reported to be potential biological antioxidants. Fullerenes possess large numbers of conjugated double bonds and low lying lowest unoccupied molecular orbitals (LUMOs) which can easily take up an electron, making an attack on radical species highly possible. It is reported that up to 34 methyl radicals have been added onto a single C_{60} molecule. This quenching process appears to be catalytic in that the fullerene can react with many superoxides without being consumed. Because of this feature, fullerenes are considered to be the world's most efficient radical scavenger and are described as radical sponges. The major advantage of using fullerenes as medical antioxidants is their ability to localize within the cell to mitochondria or other cell compartment sites, where, in diseased states, the production of free radicals takes place (Azzam and Domb 2004).

The nature of a fullerene cage as a potential "isolation chamber" indicates the possibility of carrying an unstable atom, for example, a metal atom, within the interior of the molecular cage, forming so-called endofullerenes/metallofullerenes that would be able to isolate reactive atoms from their environment. Several studies have already shown that fullerene cages are relatively non-toxic and are resistant to body metabolism. Biodistribution studies with water-soluble derivatives of C_{60} demonstrate that these compounds are primarily localized in the liver and their clearance from the body is very slow. Metallofullerenes introduce no release of the captured metal atom under *in-vivo* conditions, in contrast to metal chelates, and they show potential in diagnostic applications. Endofullerenes can be applied as magnetic resonance imaging (MRI) contrast agents, X-ray imaging agents, or radiopharmaceuticals. In another approach, fullerene derivatives were used as a carrier for serum protein profiling, which is a powerful tool for the identification of protein signatures associated with pathologies and for biomarker discovery, using the material-enhanced laser

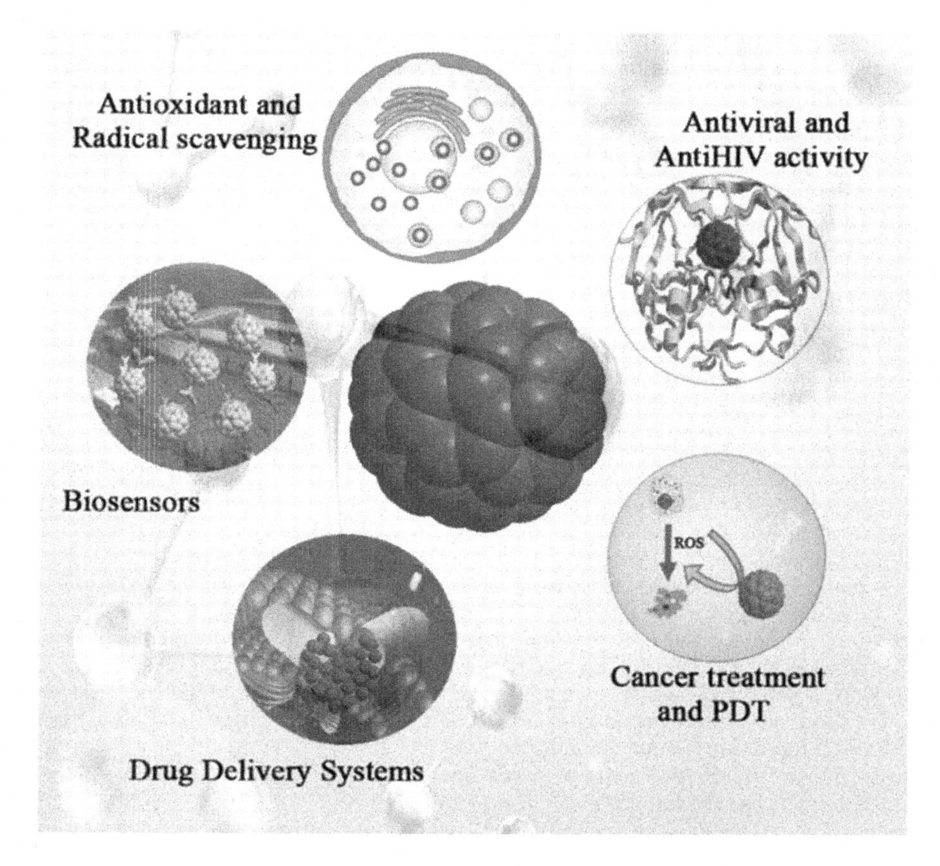

FIGURE 10.4 Various biomedical applications of fullerenes.

desorption/ionisation (MELDI) technique (Feuerstein et al. 2006). Figure 10.4 illustrates the schematic representation of various biomedical applications of fullerenes (Goodarzi et al. 2017).

10.5 CONCLUSION

The identification and extraction of natural fullerene persist as both interesting and puzzling. Fullerenes are highly resistant, metamorphic, and highly reactive. Their low abundance and scarcity are the main reason for the difficulty in extraction. Fullerenes are found only in areas where many high-energy conditions exist, like lightning, wildfire, etc., at the Cretaceous-Tertiary boundary, or meteoritic impacts. For example, lightning leads to a large variety of unexpected changes. One of the best is that lightning tends to melt the ground where it strikes. When these energy sources are active, they may produce conditions similar to those employed in the laboratory to synthesize fullerenes. Different methods were used for the extraction and confirmation of fullerene. Mass spectroscopy and HPLC are the major techniques used for this purpose. The molecular ion value *m/z* = 720 amu corresponds to C_{60} fullerene and *m/z* = 840 amu corresponds to C_{70} fullerene. Fullerenes and their derivatives offer excellent potential applications to the health care sector. In the biological field, they are reported to be potential antiviral or antioxidant agents. Photoexcitation of fullerenes is also reported by using PDI technique. In addition to this, fullerene derivatives are also used as a carrier for serum protein by using the MELDI technique.

10.6 REFERENCES

An, Y.-Z., C.-H. B. Chen, J. L. Anderson, D. S. Sigman, C. S. Foote, and Y. Rubin. "Sequence-specific modification of guanosine in DNA by a C60-linked deoxyoligonucleotide: Evidence for a non-singlet oxygen mechanism." *Tetrahedron* 52, no. 14 (1996): 5179–5189.

Azzam, T., and A. J. Domb. "Current developments in gene transfection agents." *Current Drug Delivery* 1, no. 2 (2004): 165–193.

Barceló Cullerés, D. "Occurrence of C-60 and related fullerenes in the Sava River under different hydrologic conditions." *Science of the Total Environment* 2018 (2018): 1108–1116.

Becker, L., J. L. Bada, R. E. Winans, J. E. Hunt, T. E. Bunch, and B. M. French. "Fullerenes in the 1.85-billion-year-old Sudbury impact structure." *Science* 265, no. 5172 (1994): 642–645.

Belkin, A., A. Hubler, and A. Bezryadin. "Self-assembled wiggling nano-structures and the principle of maximum entropy production." *Scientific Reports* 5, no. 1 (2015): 1–6.

Bhandari, N., M. Vairamani, A. C. Kunwar, and B. Narasaiah. "Natural fullerenes from the Cretaceous-Tertiary boundary layer at Anjar, Kutch, India." *Catastrophic Events and Mass Extinctions: Impacts and Beyond* 356 (2002): 345.

Brown, S. B., E. A. Brown, and I. Walker. "The present and future role of photodynamic therapy in cancer treatment." *The Lancet Oncology* 5, no. 8 (2004): 497–508.

Buseck, P. R. "Geological fullerenes: review and analysis." *Earth and Planetary Science Letters* 203, no. 3–4 (2002): 781–792.

Buseck, P. R., S. J. Tsipursky, and R. Hettich. "Fullerenes from the geological environment." *Science* 257, no. 5067 (1992): 215–217.

Carboni, A., R. Helmus, E. Emke, N. van den Brink, J. R. Parsons, K. Kalbitz, and P. de Voogt. "Analysis of fullerenes in soils samples collected in The Netherlands." *Environmental Pollution* 219 (2016): 47–55.

Chen, Z., P. Westerhoff, and P. Herckes. "Quantification of C60 fullerene concentrations in water." *Environmental Toxicology and Chemistry: An International Journal* 27, no. 9 (2008): 1852–1859.

Chijiwa, T., T. Arai, T. Sugai, H. Shinohara, M. Kumazawa, M. Takano, and S.-I. Kawakami. "Fullerenes found in the Permo-Triassic mass extinction period." *Geophysical Research Letters* 26, no. 6 (1999): 767–770.

Daly, T. K., P. R. Buseck, P. Williams, and C. F. Lewis. "Fullerenes from a fulgurite." *Science* 259, no. 5101 (1993): 1599–1601.

Elsila, J. E., N. P. de Leon, F. L. Plows, P. R. Buseck, and R. N. Zare. "Extracts of impact breccia samples from Sudbury, Gardnos, and Ries impact craters and the effects of aggregation on C60 detection." *Geochimica et Cosmochimica Acta* 69, no. 11 (2005): 2891–2899.

Fang, P. H., and R. Wong. "Evidence for fullerene in a coal of Yunnan, Southwestern China." *Material Research Innovations* 1, no. 2 (1997): 130–132.

Fang, P. H., F. Chen, R. Tao, B. Ji, C. Mu, E. Chen, and Y. He. "Fullerene in some coal deposits in China." In *Natural Fullerenes and Related Structures of Elemental Carbon*, pp. 257–266. Springer, Dordrecht, 2006.

Feuerstein, I., M. Najam-ul-Haq, M. Rainer, L. Trojer, R. Bakry, N. H. Aprilita, G. Stecher et al. "Material-enhanced laser desorption/ionization (MELDI): A new protein profiling tool utilizing specific carrier materials for time of flight mass spectrometric analysis." *Journal of the American Society for Mass Spectrometry* 17, no. 9 (2006): 1203–1208.

Frank, O., J. Jehlička, V. Hamplová, and A. Svatoš. "The search for fullerenes in rocks from the Ries impact crater." *Meteoritics & Planetary Science* 40, no. 2 (2005): 307–314.

Friedman, S. H., D. L. DeCamp, R. P. Sijbesma, G. Srdanov, F. Wudl, and G. L. Kenyon. "Inhibition of the HIV-1 protease by fullerene derivatives: Model building studies and experimental verification." *Journal of the American Chemical Society* 115, no. 15 (1993): 6506–6509.

Ghevariya, Z. G., and S. M. Sundaram. "K/T boundary layer in Deccan intertrappeans at Anjar, Kutch." *The Cretaceous-Tertiary Event and Other Catastrophes in Earth History* 307 (1996): 417.

Goodarzi, S., Da Ros, T., Conde, J., Sefat, F., & Mozafari, M. (2017). Fullerene: Biomedical engineers get to revisit an old friend. *Materials Today, 20*(8), 460–480.

Heymann, D., and W. S. Wolbach. "Fullerenes in the Cretaceous-Tertiary boundary." In *Natural Fullerenes and Related Structures of Elemental Carbon*, pp. 191–212. Springer, Dordrecht, 2006.

Heymann, D., L. P. F. Chibante, R. R. Brooks, W. S. Wolbach, and R. E. Smalley. "Fullerenes in the Cretaceous-Tertiary boundary layer." *Science* 265, no. 5172 (1994a): 645–647.

Heymann, D., W. S. Wolbach, L. P. F. Chibante, R. R. Brooks, and R. E. Smalley. "Search for extractable fullerenes in clays from the Cretaceous/Tertiary boundary of the Woodside Creek and Flaxbourne River sites, New Zealand." *Geochimica et cosmochimica acta* 58, no. 16 (1994b): 3531–3534.

Heymann, D., A. Korochantsev, M. A. Nazarov, and J. Smit. "Search for fullerenes C60and C70in Cretaceous–Tertiary boundary sediments from Turkmenistan, Kazakhstan, Georgia, Austria, and Denmark." *Cretaceous Research* 17, no. 3 (1996): 367–380.

Heymann, D., B. O. Dressler, J. Knell, M. H. Thiemens, P. R. Buseck, R. B. Dunbar, and D. Mucciarone. "Origin of carbonaceous matter, fullerenes, and elemental sulfur in rocks of the Whitewater Group, Sudbury impact structure, Ontario, Canada." *Special Paper of the Geological Society of America* 339 (1999): 345–360.

Heymann, D., L. W. Jenneskens, J. Jehlička, C. Koper, and E. Vlietstra. "Fullerenes in extracts of impact breccia samples from Sudbury, Gardnos, and Ries impact craters and the effects of aggregation on C60 detection." *Fuller Nanotub Car* 11 (2003): 333–370.

Jehlička, J., A. Svatoš, O. Frank, and F. Uhlik. "Evidence for fullerenes in solid bitumen from pillow lavas of Proterozoic age from Mitov (Bohemian Massif, Czech Republic)." *Geochimica et Cosmochimica Acta* 67, no. 8 (2003): 1495–1506.

Krusic, P. J., E. Wasserman, P. N. Keizer, J. R. Morton, and K. F. Preston. "Radical reactions of C60." *Science* 254, no. 5035 (1991): 1183–1185.

Lalwani, G., and B. Sitharaman. "Multifunctional fullerene-and metallofullerene-based nanobiomaterials." *Nano Life* 3, no. 3 (2013): 1342003.

Misra, K. S., M. R. Hammond, A. V. Phadke, F. Plows, U. S. N. Reddy, I. V. Reddy, G. Parthasarathy, C. R. M. Rao, B. N. Gohain, and D. Gupta. "Occurrence of fullerene bearing Shungite Suite Rock in Mangampeta Area, Cuddapah District, Andhra Pradesh." *Journal of Geological Society of India* (Online archive from Vol 1 to Vol 78) 69, no. 1 (2007): 25–28.

Mosin, O. V., and I. Ignatov. "The structure and composition of natural carbonaceous fullerene containing mineral shungite." *International Journal of Advanced Scientific and Technical Research* 6, no. 11–12 (2013): 9–21.
Mossman, D., G. Eigendorf, D. Tokaryk, F. Gauthier–Lafaye, K. D. Guckert, V. Melezhik, and C. E. Farrow. "Testing for fullerenes in geologic materials: Oklo carbonaceous substances, Karelian shungites, Sudbury Black Tuff." *Geology* 31, no. 3 (2003): 255–258.
Parthasarathy, G., R. Srinivasan, M. Vairamani, K. Ravikumar, and A. C. Kunwar. "Fullerene from the proterozoic shungite deposit at Kondopoga, Karelia, Russia: Isotopic and spectroscopic studies." *Chinese Science Bulletin* 43, no. S-1 (1998): 98–98.
Poreda, R. J., and L. Becker. "Fullerenes and interplanetary dust at the Permian-Triassic boundary." *Astrobiology* 3, no. 1 (2003): 75–90.
Qiao, R., A. P. Roberts, A. S. Mount, S. J. Klaine, and P. C. Ke. "Translocation of C60 and its derivatives across a lipid bilayer." *Nano Letters* 7, no. 3 (2007): 614–619.
Sanchís, J., D. Božović, N. A. Al-Harbi, L. F. Silva, M. Farré, and D. Barceló. "Quantitative trace analysis of fullerenes in river sediment from Spain and soils from Saudi Arabia." *Analytical and Bioanalytical Chemistry* 405, no. 18 (2013): 5915–5923.
Santos, V. G., M. Fasciotti, M. A. Pudenzi, C. F. Klitzke, H. L. Nascimento, R. C. L. Pereira, W. L. Bastos, and M. N. Eberlin. "Fullerenes in asphaltenes and other carbonaceous materials: natural constituents or laser artifacts." *Analyst* 141, no. 9 (2016): 2767–2773.
Shareef, A., G. Li, and R. S. Kookana. "Quantitative determination of fullerene (C60) in soils by high performance liquid chromatography and accelerated solvent extraction technique." *Environmental Chemistry* 7, no. 3 (2010): 292–297.
Sijbesma, R., G. Srdanov, F. Wudl, J. A. Castoro, C. Wilkins, S. H. Friedman, D. L. DeCamp, and G. L. Kenyon. "Synthesis of a fullerene derivative for the inhibition of HIV enzymes." *Journal of the American Chemical Society* 115, no. 15 (1993): 6510–6512.
Volkova, I. B., and M. V. Bogdanova. "Petrology and genesis of Karelian shungite: high rank coal." *International Journal of Coal Geology* 6, no. 4 (1986): 369–379.
Zhang, Y., S. Sadjadi, and C.-H. Hsia. "Hydrogenated fullerenes (fulleranes) in space." *Astrophysics and Space Science* 365, no. 4 (2020): 1–8.

Index

C

D

H

I

T

U

V

W

X

Y

Z